UNEVEN CONNECTIONS

A PARTIAL HISTORY OF THE MOBILE PHONE IN PAPUA NEW GUINEA

UNEVEN CONNECTIONS

A PARTIAL HISTORY OF THE MOBILE PHONE IN PAPUA NEW GUINEA

ROBERT J. FOSTER

Australian
National
University

ANU PRESS

PACIFIC SERIES

*I have been lucky to have the unconditional support of Nancy Fried Foster
in everything connected with the making of this book
and with everything else that really matters.
I dedicate this book to Nancy with love and gratitude.*

Australian
National
University

ANU PRESS

Published by ANU Press
The Australian National University
Canberra ACT 2600, Australia
Email: anupress@anu.edu.au

Available to download for free at press.anu.edu.au

ISBN (print): 9781760466251
ISBN (online): 9781760466268

WorldCat (print): 1411589617
WorldCat (online): 1411588939

DOI: 10.22459/UC.2024

Cover design and layout by ANU Press.

Cover photograph: Woven string bag (wool-based yarn) made by women of the Mt Hagen Handicraft Group, Papua New Guinea. Nokia mobile phone model 1616-2: RH 125, circa 2010. Photo by N. Foster.

This book is published under the aegis of the Pacific editorial board of ANU Press.

Publication of this book has been supported by the ANU vice-chancellor's strategic funds for flagship titles at ANU Press.

Contents

List of Illustrations   ix

Acknowledgements   xi

Introduction: The Moral Economy of Mobile Phones
in Papua New Guinea   1

Part I. After Liberalisation: An Evolving Mobile Market

1. The Politics of Mobile Phone Infrastructure: Licences, Towers
   and Gateways   23

2. Making a (National) Market: Advertisements, Promotions
   and Sponsorships   57

Part II. Consumer Uptake: Freedom and Constraint

3. Mobile Economies: Prepaid, Gift and Informal   79

4. Smartphones and Data: Convergence and Content   109

Part III. Delimiting Mobility: Regulation and Responsibility

5. Mobile Disruption: Regulation, Surveillance and Censorship   133

6. Connecting the Unconnected: Corporate Social Responsibility
   and Post-Political Governance   161

Conclusion: Infrastructural Citizenship and Uneven Connections   183

References   191

List of Illustrations

Figures

Figure I.1. Digicel advertising at the Goroka marketplace. 3

Figure 1.1. bmobile-Vodafone headquarters, Port Moresby, 2017. 28

Figure 1.2. Cover image of a GSMA report. 35

Figure 2.1. Digicel newspaper ad (Western Highlands coverage). 61

Figure 2.2. Digicel newspaper ad (East New Britain coverage). 61

Figure 2.3. Telikom newspaper ad (Independence Day). 62

Figure 2.4. Digicel newspaper ad (Swap SIM cards). 63

Figure 2.5. Digicel newspaper ad (11 toea talk). 63

Figure 2.6. Digicel newspaper ad (Traditional dress Independence Day). 64

Figure 2.7. Digicel newspaper ad ('Call Me' service). 64

Figure 2.8. bemobile newspaper ad (bemobile Orange Men). 65

Figure 2.9. bemobile newspaper ad (19 kina Christmas Eve handset). 67

Figure 2.10. Digicel newspaper ad (2 for 1 handset offer). 67

Figure 2.11. Digicel newspaper ad (Thank You 10 kina free credit). 68

Figure 2.12. bemobile newspaper ad (Win a pig contest). 70

Figure 2.13. Digicel newspaper ad (Hiri Moale sponsorship). 74

Figure 3.1. Digicel 5 kina scratch-off voucher or 'flex card'. 81

Figure 3.2. Digicel text message for a credit loan. 90

Figure 3.3. Digicel newspaper ad (Street Vendor Programme). 101

Figure 3.4. Digicel newspaper ad (Larry Koavea, My life is better). 101

Figure 3.5. Top-up transaction, Goroka, 2015. 102

Figure 3.6. Mobile phone repair technician's table, Goroka, 2015.   103

Figure 5.1. Telikom PNG newspaper ad (Fixed Line).   145

Figure 6.1. Ad for PNG Power's Easipay service at bmobile-Vodafone retail store, Port Moresby, 2018.   171

Figure 6.2. Westpac instore banking at urban settlement near Port Moresby, 2016.   174

Maps

Map I.1. Country map of Papua New Guinea with provincial borders and capitals.   2

Map 1.1. Digicel coverage map, circa 2015.   37

Map 1.2. bmobile 2022 coverage map with 2G, 3G and 4G zones.   47

Acknowledgements

I owe thanks to a legion of individuals and institutions who in ways big and small facilitated, assisted and encouraged the drawn-out research process that led finally to this book. I ask forgiveness from anyone whom I have inadvertently but inevitably failed to acknowledge here.

An Australian Research Council (ARC) Discovery Project Grant (DP140103773, *The Moral and Cultural Economy of Mobile Phones in the Pacific*) awarded to Heather A Horst (CI) and myself (PI) provided material support. Additional material support came from the University of Rochester (School of Arts & Sciences) in the form of academic leave and research funds associated with my appointment as Richard L Turner Professor of Humanities.

Research in Papua New Guinea (PNG) was conducted under the auspices of the PNG National Research Institute (NRI) and in collaboration with the Department of Anthropology and Sociology at the University of Papua New Guinea (UPNG) in Port Moresby and the Centre for Social and Creative Media (CSCM) at the University of Goroka. I thank Georgia Kaipu of NRI for arranging research permission and visas, and Dr linus digim'Rina (UPNG) and Dr Verena Thomas (formerly CSCM) for aid in coordinating the research in Port Moresby and Goroka.

I especially thank my research partner Professor Heather A Horst for her steady friendship, keen insights and organisational skills in managing our ARC grant. I have benefited from the affiliations that Professor Horst arranged for me first at the Digital Ethnography Research Centre (DERC), RMIT University and then at the Institute for Culture and Society, Western Sydney University. I also thank Dr Amanda HA Watson for her generous collegiality in sharing resources and contacts and in offering welcome hospitality in Port Moresby. Dr Watson's help and advice were instrumental in setting my research plans in motion.

Melbourne: I thank the staff and members of DERC for making my time as a visiting fellow in 2015 stimulating and enjoyable: John Postill, Larissa Hjorth, Tania Lewis, Supriya Singh and Sarah Pink. Luke Gaspard and Christine Schmidt assisted in archiving newspaper articles and exploring website designs. Several Melbourne-based colleagues offered warm hospitality and engaging conversation: Martha Macintyre, John Cox and Elizabeth Bonshek.

Port Moresby: Dr linus digim'Rina arranged the research assistance of Jason Kariwiga and Alex Nava as well as the taxi services of Barry Kapik. Librarians Paul Oerepa Jagipa and Percy Roary made NRI's rare print collection of old PNG newspapers available to me. Anton Neinaka provided logistical support on visits in 2017 and 2018. Paul Barker (Institute of National Affairs) shared his account of Digicel's arrival in PNG. Many other colleagues contributed informative histories, personal anecdotes and introductions to industry insiders. My thanks to Andrew Moutu, Denis Crowdy, Nicolas Garnier, Don Niles, Martyn Namorong, Emmanuel Narokobi, Olga Temple and Alan Robson. I am indebted to several Tanga Islanders living in Port Moresby for accepting me into their community, especially the late Cletus Ngaffkin, Bernard Soari, Lorraine Maro, Andreas Neansugel, Lynn Funmat Kotong, the late Erwin Maritua, and Felix Nebanat.

Many corporate, government and non-government organisation officials consented to interviews on various aspects of the telecommunications industry. I am grateful to all of them for fitting me into their busy schedules. It should be noted that this book does not necessarily reflect the official policy or position of any of these individuals or of the organisations that employed these individuals at the time of the research.

I thank the following individuals at the National Information and Communications Technology Authority (NICTA) for their time and attention: Charles Punaha, Jackson Kariko, Gibson Tito, Hans Adeg, Vlado Doncevski and Andirauga Paru Nongkas. Ilikomau Ali, Delly Morofa and Cyril Kruak welcomed my questions at the Office of Censorship, as did Julie Wulwarau at the PNG Independent Consumer and Competition Commission (ICCC).

I learned much about the state of mobile banking and microfinance from Tony Westaway of MiBank, Sweta Sud of the PNG Institute of Banking & Business Management and Saliya Ranasinghe of the Bank of PNG's Microfinance Expansion Project. Raphael Waiyalaka kindly allowed me to accompany him on site visits to retail agents of Westpac's Everywhere Banking. Simon Schwall of BIMA helpfully explained his company's mobile microinsurance business, while Andrew Johnston of Pacificview Multimedia updated me about the effects of digital technologies on the advertising industry in PNG.

Several Digicel PNG officials who have since left the company were generous in meeting with me and sharing their views of the mobile mediascape: John Mangos, Gary Seddon, Genevieve Daniels, Shivan Bhargava and Andrew Hodgson. I thank Beatrice Mahuru and Jennifer McConnell, formerly of the Digicel PNG Foundation, for their insights and the opportunity to sit in on a foundation-sponsored Men of Honour event.

Michael Donnelly at Telikom PNG and Sundar Ramamurthy at bmobile-Vodafone provided candid C-suite overviews of mobile communications in PNG. Gordon Maitava at Telikom PNG and Tidman Ikosi at bmobile-Vodafone fielded requests for information. Bhanu Sud helpfully discussed the acquisition of EMTV by Telikom and the future of media convergence. The late Peter Loko shared the story of his experience as CEO of Telikom PNG at the time of Digicel's launch. Tony Morisause of PNG DataCo supplied an engineer's perspective on the challenges of telecommunications infrastructure.

Goroka: Dr Verena Thomas (CSCM) co-supervised Wendy Bai Magea, a BA Honours student at the University of Goroka, sponsored by our ARC grant, whose research underpinned the production of a short video (*Mobail Goroka*) on mobile airtime vendors. Thanks go to all involved in producing the video, especially the director Dr Jackie Kauli.

Wendy Bai Magea, Alessandra Mel and Ben Ruli provided essential assistance with interviews. Milan Boie and Dilen Doiki assisted with photography. Fraser Macdonald added useful advice.

Joe Elua and Josephine Ketauwo at Telikom PNG provided a view of the company's operations in the Goroka area. Sammy Sari at the Goroka office of the ICCC discussed consumer concerns with me. I am particularly indebted to Pattsie Gotaha for taking me on illuminating site visits to agents of Australia and New Zealand Banking Group (ANZ) goMoney. I owe an even greater debt to Michael Agua and Luke Natapol, who brought me inside the informal economy of mobile phone repair and sale of airtime vouchers, and who, along with Joseph Kaupa, made possible the research of Wendy Bai Magea and the video that features their work.

George Anian of Coffee Connections and Bill Humphrey of Coffee Industry Corporation answered my questions about the use of mobile phones in coffee farming. My visits to Goroka fortunately coincided with those of friend and colleague Mark Busse, who introduced me to the workings of the Goroka marketplace. Olivia Barnett-Naghshineh and Christopher Little also shared insights from their ethnographic research on commerce in the town.

Many thanks to Ennie Gasowo and the staff of the Lutheran Guest House for providing comfortable and convenient accommodations.

Suva: Although this book focuses on PNG, it is informed by comparative research in Fiji led by Professor Horst. Sandra Tarte arranged affiliation with the University of the South Pacific (USP); Karen Brison offered sound advice and a pivotal introduction. Several officers of Digicel Fiji and Digicel Pacific shared their previous experiences working in PNG: Darren McLean, Awais Malik and Peter Rigamoto. Katie Taylor was exceptionally generous in explaining the history of Digicel's marketing campaigns in the Pacific Islands region.

Conversations with several colleagues at USP enhanced my understanding of mobile phone use in the region: Romitesh Kant, Jope Tarai, Jason Titifanue and Glen Finau. Marc Lipton and the late Donnie Defreitas of the USP – Pacific ICT Regulatory Resource Centre shared their extensive expertise on ICT (information and communications technology) service provision and the role of regulators in the Pacific.

In Suva, Professor Horst and I enjoyed the warm hospitality of Racheal Bale Spencer and Mike Spencer, and John Cox and Georgina Phillips. *Vinaka!*

Sydney: During the long span of this project, I have passed through Sydney many times, staying at the Redfern home of my archaeologist colleagues Robin Torrence and Peter White. Their unselfish hospitality and engaging conversation were always restorative. I value the friendship we have developed over the years as a happy byproduct of this project.

Erin Taylor contributed to the conceptualisation of the project at a formative stage. Other colleagues in Sydney have helped this book along with their comments, suggestions and enthusiasm: Gerard Goggin, Gay Hawkins, Malini Sur, Rose Lilley, Neil Maclean and Dick Bryan.

Canberra: I am grateful to Scott MacWilliam for his critical perspective and genuine solidarity, each equally uncompromising. I am also grateful for the intellectual provocation and good company of Margaret Jolly, who invited me to speak at the 'Worlding Oceania' symposium in 2015 at The Australian National University (ANU). These two dear friends have shaped how I imagine the breadth and diversity of contemporary Pacific studies.

Thanks to Sarah Logan for inviting me to a workshop on Information and Communication Technologies in Melanesia sponsored by the ANU State, Society & Governance Program, and to Alan Rumsey and Francesca Merlan for their interest in this research and hospitality in Canberra.

I thank Stewart Firth for accepting this book in the ANU Press Pacific Series and Sarah Sky for her work as production editor. I am thrilled that ANU Press will make this book available as an open-access publication, allowing Pacific Islanders, including people in PNG, to read it on their mobile phones. Two anonymous reviewers supplied constructive feedback, and Beth Battrick copyedited the manuscript with diligent care.

Rochester: The Department of Anthropology at the University of Rochester has supported me in multiple ways. Ro Ferreri and Donna Mero provided the requisite administrative support professionally and without fail. My faculty colleagues assumed teaching and service responsibilities while I was on academic leave, which Dean Gloria Culver made possible. David Lipset has been a valuable email commentator on my writing about mobile phones in PNG.

Students in the Graduate Program in Visual and Cultural Studies at the University of Rochester provided research assistance over the course of the project: Kelly Long, Lina Zigelyte, Sara Collins, Alana Wolf-Johnson and Daisuke Kawahara.

***

Portions of Chapter 1 and Chapter 5 appeared in: Foster, Robert J. 2020. 'The Politics of Media Infrastructure: Mobile Phones and Emergent Forms of Public Communication in Papua New Guinea'. *Oceania* 90, no. 1: 18–39 (doi.org/10.1002/ocea.5241).

A portion of Chapter 3 appeared in: Foster, Robert J. 2018. 'Top Up: The Moral Economy of Prepaid Mobile Phone Subscriptions'. In *The Moral Economy of Mobile Phones: Pacific Islands Perspectives,* edited by Robert J Foster and Heather A Horst, 107–125. Canberra: ANU Press (doi.org/10.22459/MEMP.05.2018.06).

A small portion of Chapter 6 appeared in: Foster, Robert J. 2017. 'The Corporation in Anthropology'. In *The Corporation: A Critical, Multi-Disciplinary Handbook*, edited by Grietje Baars and Andre Spicer, 111–33. Cambridge: Cambridge University Press (doi.org/10.1017/9781139681025.006).

A portion of the Conclusion appeared in: Foster, Robert J. 2023. 'Tenuous Connectivity: Time, Citizenship and Infrastructure in a Papua New Guinea Telecommunications Network'. *The Asia Pacific Journal of Anthropology* 24, no. 2: 94–115 (doi.org/10.1080/14442213.2023.2177330).

Introduction: The Moral Economy of Mobile Phones in Papua New Guinea

In July 2022, Digicel Pacific, a unit of the firm that brought a sea change in mobile communications to many Pacific Islands nations, was formally acquired by a subsidiary of the Australian telecommunications company Telstra. Digicel Group Ltd's founder and principal owner Denis O'Brien eulogised his legacy:

> Having established our Pacific operations as a business start-up in 2005, we depart with enormous pride in a team that has made affordable best-in-class communications available to more than 10 million people across six of the most exciting economies in the region. (Brennan 2022)[1]

Digicel had indeed radically transformed the regional mediascape. In Papua New Guinea (PNG), Digicel's largest and most profitable Pacific market, the sale marked the end of an epic chapter in the fraught history of 'uniting a nation' through telecommunications services (Sinclair 1984).

Digicel launched operations in PNG in 2007. One year earlier, there were an estimated 100,000 mobile cellular subscriptions in the country, or about 2 per 100 inhabitants. Three years later in 2010, there were an estimated 1.9 million subscriptions or 26 per 100 inhabitants. When I returned to PNG that year after an eight-year hiatus, evidence of the transformation in mobile telecommunications in the capital city of Port Moresby was plainly visible. Mobile phones, like Digicel billboards, seemingly appeared

---

1   As a wholly owned subsidiary of Digicel Group Ltd, Digicel Pacific operated in six countries: Fiji, Nauru, Tonga, Samoa, Vanuatu and PNG. Telstra planned to continue trading under the Digicel brand name (Business Advantage PNG 2021).

everywhere. I purchased a Digicel subscriber identity module or SIM card for my old Siemens C60 handset and joined the conversation. When I returned in 2012, I upgraded to a Motorola WX181, a classic 'candy bar' phone with 1.45-inch colour display, LED torch light and FM radio. By then, the number of mobile subscriptions had grown to an estimated 2.7 million or 35 per 100 Papua New Guineans, the overwhelming majority of whom communicated on Digicel's network. A future of similarly remarkable growth in the number of mobile phone users looked all but inevitable.[2]

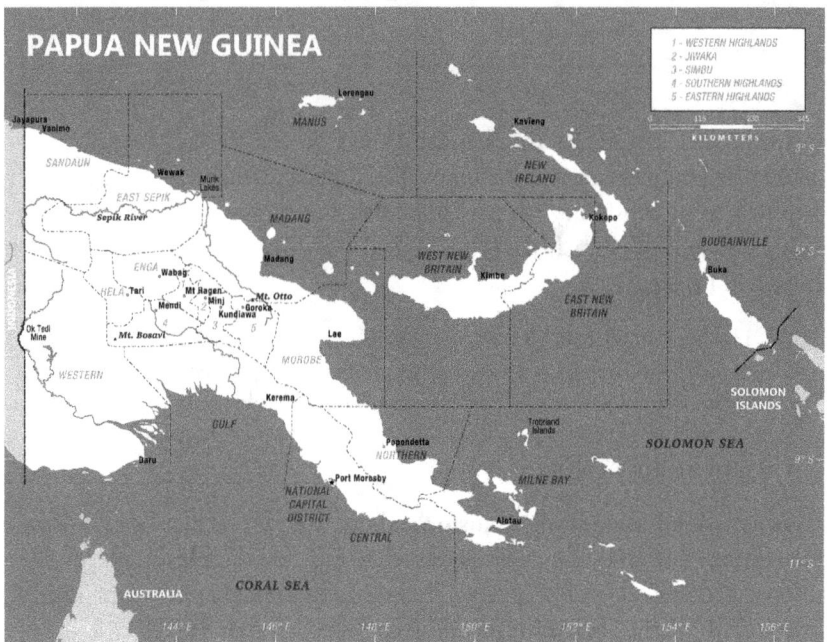

Map I.1. Country map of Papua New Guinea with provincial borders and capitals.

Source: The Map Shop.

---

2   For statistics of mobile subscriptions in PNG, see International Telecommunication Union (ITU) World Telecommunication/ICT Indicators Database, data.worldbank.org/indicator/IT.CEL.SETS.P2? end=2020&locations=PG&start=1960&view=chart, accessed 9 December 2022.

Digicel's Pacific Moment

Figure I.1. Digicel advertising at the Goroka marketplace.
Source: Photo by D Doiki.

By all accounts, the story of the so-called mobile revolution in PNG begins with the arrival of Digicel Group Ltd and its affiliates (see e.g. Cave 2012). Founded by Irish businessman Denis O'Brien, the nimble mobile network provider is privately owned—at one point 94 per cent by O'Brien himself—and registered in Bermuda with headquarters in Kingston, Jamaica. According to the *Irish Times*, O'Brien 'extracted at least $1.9 billion (€1.7 billion) of dividends from Digicel between 2007 and 2014' (Brennan 2019). In 2015, *Forbes* listed O'Brien, who resides in Malta for tax purposes, as one of the world's top 200 billionaires. *Forbes* estimated O'Brien's worth at USD3.5 billion in December 2022 (Forbes n.d.).[3]

Digicel started operations in Jamaica in 2001 and expanded throughout the Caribbean and Central America over the next five years (Horst and Miller 2006). By 2016, Digicel's global mobile subscriber base had grown from

---

3   For details of O'Brien's early career and business interests in media and communications, see Creaton (2010).

400,000 in 2002 to 13.6 million across more than 30 countries. Following the sale of Digicel Pacific in 2022, Digicel Group continued to operate in 25 Caribbean and Central American countries, claiming more than 10 million customers.

Digicel strategically targeted 'high risk' countries with fairly small populations in the developing world, introducing aggressive competition into markets where telecommunications services were provided, often ineffectively, by state-sponsored monopolies. Digicel's interest in the Pacific Islands officially began with the award of a licence to operate in Samoa in 2006 and, soon afterwards, in PNG, Fiji, Tonga, Vanuatu and Nauru. Digicel also attempted to obtain a licence in Solomon Islands in 2007 but was blocked by a local effort to protect the state-owned incumbent, Solomon Telekom. After subsequent attempts to acquire a licence in 2008 and 2009, Digicel lost its bid to its only competitor in PNG, Bemobile, then co-owned by the PNG state and a consortium of private and public investors.

During the first five years of operations, Digicel managed to build goodwill and enjoy popularity among consumers across the Pacific region. Competition reduced the price of calls and increased the availability and affordability of handsets. The establishment of mobile networks in poorly served areas enabled Digicel to reach marginalised segments of the population and to build a base of low-income consumers.[4] Digicel's self-proclaimed 'bigger and better network' allowed for stronger connections between family and friends in remote rural villages and growing urban centres.

Digicel also engaged in well-publicised acts of corporate philanthropy, such as providing aid after cyclones. The Digicel Foundation launched in PNG in 2008, supporting community initiatives to improve health, education and the welfare of women and youths (Chapter 6). At the same time, the company invested in the cultural life of its new consumer markets by sponsoring rugby and other popular sports teams as well as music competitions. Digicel embraced digital media convergence, transforming into a provider not only of connectivity, but also of content such as news, sports and lifestyle stories (Chapter 4). The company also diversified into the consumer entertainment market, setting up its own television network and acquiring cable and satellite television broadcasters and internet service providers across the Pacific region.

---

4   Digicel also offered a host of products and services to business customers (known as B2B), consideration of which falls beyond the scope of this book.

Digicel Pacific's adjusted annual earnings before interest, tax, depreciation and amortisation (EBITDA) was approximately USD222 million in the most recent financial year before its sale to Telstra. It employed 1,700 staff members. PNG was by far the company's largest and most lucrative market, where Digicel held up to 97 per cent of mobile market share (Digicel Group Ltd 2015). According to McLeod (2020):

> in the year to March 2019, PNG operations brought in revenue of more than US$340 million and made a contribution to group earnings (EBITDA) of $160.5 million. PNG's contribution was worth 16.7 per cent of total EBITDA and was the largest from a single market within the group. Revenue and profitability increased even as the company lost a half million mobile subscribers, which it says was the result of government policies requiring the registration of SIM cards.

By contrast, Fiji—where Digicel held an approximately 33 per cent market share (Digicel Group Ltd 2015; the actual rate might have been as low as 20 per cent)—was one of its least successful operations. Vodafone Fiji dominated the market due to its historical relationship with Fijian state agencies as well as the goodwill associated with its philanthropic foundation (see Horst 2018). Fiji's skilled workforce and relatively comfortable standard of living, however, made it an ideal hub for some of Digicel Pacific's regional marketing and legal operations.

Digicel's departure from the Pacific invites a review of what happened in PNG after the liberalisation of the telecommunications sector in 2005 made competition in the market for mobile communications possible. This story deserves telling, even if an exhaustive version is beyond the means of a short book. By the time Digicel left PNG, the presence of mobile phones had become normalised for a large part of the country's urban population.[5] As in many other developing countries, public pay phones had disappeared while fixed lines remained scarce, and a generation of youth could recall growing up with only mobile telephony. How, then, did the so-called mobile revolution unfold? Who were its agents and beneficiaries? What were its effects and how were these effects distributed?

---

5   There were an estimated 55 mobile cellular subscriptions per 100 people in PNG in 2020, according to ITU statistics: data.worldbank.org/indicator/IT.CEL.SETS.P2?end=2020&locations=PG& start=1960&view=chart, accessed 22 December 2022.

This book addresses these broad questions through a selective history of the mobile phone in PNG, one that documents aspects of a socio-material transformation in the lives of many Papua New Guineans for whom access to affordable telecommunications was a warmly welcome novelty. In so doing, the book contributes to the current anthropology of mobile phones in Melanesia (Hobbis 2019) and elsewhere (Horst 2021), especially the Global South. The approach taken extends previous work that describes a moral economy of mobile phones in which companies, consumers and state agents constantly negotiate who owes what to whom (Foster and Horst 2018). My aim is to demonstrate how these negotiations resulted in outcomes that often confounded the expectations of policymakers and ordinary citizens, and posed challenges to companies, consumers and state agents alike in establishing and experiencing new forms of connectivity. The uptake of mobile phones in PNG did not proceed in ways that can be called typical in any sense; it happened in circumstances—historical, economic, political and cultural—that were specific to PNG. But by telling this story with a guiding concern for the relations between and among consumers, companies and state agents, the details become relevant for a more general understanding of mobile communication in the Global South (Ling and Horst 2011; Tenhunen 2018).

Freedom, Constraint and the Anthropology of Mobile Phones

Ethnographic research on mobile phones has often focused—rightly, if perhaps predictably—on how users have readily adopted and adapted the technology for both old and new purposes. This creative appropriation extends from the ways in which users in the Global South quickly took to 'flashing' or 'beeping' with basic handsets (Donner 2008) in order to send messages without paying for airtime to the ways in which more affluent users in the Global North craft their smartphones and perforce themselves by adding and maintaining apps, email, photos and so forth that expand the capacity for communication. Daniel Miller et al. (2021: 7), for example, introduce an important book on 'the global smartphone' by noting that their 'volume is replete with amazing invention, design and application by people integrating smartphones within their everyday lives'.

A user-centred focus on creativity is valuable for countering reductive arguments about technological determination—a hallmark of the earliest (e.g. Horst and Miller 2006) and latest (e.g. Tenhunen 2018) ethnography of mobile phone use in the Global South.[6] Documenting the unanticipated uses to which mobile devices have been put similarly exposes how users as well as designers effectively construct the mobile phone in practice. Looking at what consumers do with mobile phones has so far been the strength of the emerging anthropological literature based on ethnographic research in Melanesia (see, for example, Andersen 2013; Lipset 2013; Foster and Horst 2018; Jorgensen 2014; Telban and Vávrová 2014; Kraemer 2015; Macdonald and Kirami 2015; Taylor 2015; Hobbis 2020). Ethnography demonstrates, for example, how consumers navigate the possibilities and perils of novel forms of interpersonal intimacy at the same time that they purchase and exchange airtime (phone credit) with expectations and ideals derived from longstanding practices of reciprocity and kinship.

A user-centred focus on creativity, however, requires supplementing with a complementary focus on the various constraints that users, especially in the Global South, must negotiate in order to operate their devices, creatively or otherwise. Some of these constraints are irreducibly material. For example, Donner (2015) has called attention to the 'metered mindset' of mobile users who struggle to manage their data consumption on limited budgets, while Hobbis (2020) has highlighted how environmental elements such as sand and humidity challenge the capacity of users to keep their mobile devices operable. In PNG, uncertain access to electricity makes the task of keeping a mobile device's battery continuously charged, one that can hardly be taken for granted.

The double focus that I am suggesting here raises the broader question of how an anthropology of mobile phones can bring both freedom and individual agency, on the one hand, and constraint and social structure, on the other, into the same frame of analysis. User-centred ethnography has often engaged this question by emphasising the ambivalence with which users operate their mobile phones. This theme of ambivalence is particularly pronounced in studies of users new to mobile communications in the Global South. Wallis (2013), for example, demonstrates how mobile phones enable young rural-to-urban women in China to perform a 'modern identity', but also subject these migrant workers to new forms of surveillance and exploitation by their employers. The gender, age and class of migrant women—in short, their social position—constrain their engagement with

6   Tenhunen (2018) provides a brief review of ethnographic research on mobile phones in the Global South and emphasises how such studies challenge the optimism of much of the discourse surrounding the use of mobile technologies in economic development strategies (see Chapter 3).

mobile technologies that nevertheless afford a kind of 'immobile mobility' in which phone use helps overcome the constraints of long work hours, little time off and confined social worlds (cf. Tenhunen 2018).

The earliest accounts of mobile phone use in PNG (Sullivan 2010a; Watson 2011; Lipset 2013) also foreground the ambivalence surrounding the devices. This ambivalence, however, is usually expressed in moral terms. Almost all users observed how the phones were both good and bad—enabling people to fulfill their kinship obligations to each other more readily, but equally enabling (or even creating) forms of immoral behaviour such as arranging adulterous liaisons or organising criminal activities. In this sense, mobile phones not only endow users with an enhanced agency for realising familiar cultural values, but also challenge users to constrain themselves—that is, to operate their devices in a way consistent with prevailing moral norms.

While some aspects of the dynamics of freedom and constraint as lived experience are well captured through a user-centred ethnographic approach, other aspects of this dynamic are not. As Miller et al. (2021: 21) readily admit, their ethnography of users leaves little room for considering how other forces shape the circumstances of constraint in which all mobile phone use takes place. These forces include both corporate and state actors—company executives and government regulators, for example—as well as the whole infrastructural assemblage of which the mobile handset is merely one piece. Corporations attempt to stimulate mobile usage through marketing; governments attempt to constrain mobile usage through legislation about, for example, online speech. And the presence or absence of a functioning cell tower might determine whether mobile usage is even possible in the first place. How, then, can an anthropology of mobile phones consider these forces and actors—or, at least, *some* of these forces and actors—in relation to the dynamics of freedom and constraint routinely experienced by mobile phone users?

A Moral Economy Framework

The moral economy framework developed in this book is meant first of all to provide a view of the socio-material network that mobile phones brought into being in PNG, attendant upon the government's decision to liberalise the mobile communications market in 2005 (Chapter 1). This network includes, at a minimum, the consumers who use the phones, the companies that supply consumers with handsets and airtime, and the state agents that regulate (even if haphazardly) both telecommunications companies and the infrastructure that subtends commercial mobile communications. It is the effects of all the elements in this network—including the phones themselves as material objects with definite qualities—that at any given moment inform what a user-centred ethnographic approach can tell us about mobile phones in Melanesia or elsewhere. Put differently, while it is certainly important not to reduce phone usage to the automatic outcome of either corporate strategies or state policies, it is equally important to acknowledge from the get-go that such usage occurs in circumstances that users themselves have not chosen.

The moral economy of mobile phones accordingly implies a field of shifting relations among *consumers*, *companies* and *state agents*, all of whom have their own ideas about what is good, proper and just. These ideas inform the ways in which, for example: consumers acquire and use mobile phones; companies market and price voice, text messaging and data subscriptions; and state agents regulate both the everyday use of mobile phones and the market activity of licensed competitors. Ambiguity, disagreement and ongoing negotiation about who owes what to whom are thus integral features of the moral economy of mobile phones.

Companies, consumers and state agents impinge upon each other in ways that make it difficult to consider one apart from the others. Telecommunications companies may not operate without state-issued spectrum licences; consumers cannot operate their mobile phones without registering their SIM cards, a requirement that states oblige companies to enforce; and states count on the revenue of taxes and fees collected from companies. The perspective taken here is thus intended to illuminate how companies, consumers and state agents both enable and limit each other's goals.[7] From this perspective, the mobile phone appears as a device that allows users to expand their communicative possibilities; companies to generate profits; and state agents to deliver long-awaited telecommunications services to citizens. But users, companies and state agents all must navigate an evolving series of material, fiscal, political and legal conditions in order to achieve their aims. Put differently, users, companies and state agents must all sort and deal with their claims on each other, producing and reproducing in the process a moral economy of mobile phones. It is this process of sorting

---

7   The scope of inquiry in this book is most similar to that of Doron and Jeffrey's more detailed *The Great India Phone Book* (2013), an excellent historical and anthropological collaboration that examines the roles of companies, consumers and state agents in making the mobile media landscape in India.

and dealing and this resultant moral economy, as they took shape in PNG during the period of Digicel's operation between 2007 and 2022, that this book seeks to describe and understand.

Research Background

This book is one outcome of a multi-year research project jointly undertaken with Professor Heather A Horst of Western Sydney University and funded primarily by the Australian Research Council (DP140103773). The project, titled 'The Moral and Cultural Economy of Mobile Phones in the Pacific', was designed as a comparative study of relations among consumers, companies and state agents in Fiji (Digicel's least successful market worldwide) and PNG (Digicel's largest Pacific market). Field research began in 2014 and continued until 2018; during that period, I made eight trips to PNG either by myself or with Horst, and four trips to Fiji, all with Horst. Field research involved a mix of historical and ethnographic methods, such as reading local media accounts of Digicel's arrival in Fiji and PNG and interviewing a range of corporate and government officials as well as mobile phone users (see below). Developments around mobile phones in PNG since 2018 have been tracked through social media reports, business news and relevant academic secondary literature. When the project was conceived in 2012, Horst and I presumed that the object at the centre of our study would be what we now refer to as a basic or simple (or 'one-bang') mobile phone used principally for making voice calls and sending text messages. As the project moved forward, it became clear that we needed to attend to the spread of smartphones and their use by many people as the primary and often only means for access to the internet (see Chapter 4).

This book builds on research conducted in and on PNG mainly at two different sites: the national capital of Port Moresby (population about 400,000) and the town of Goroka (population about 25,000), capital of PNG's Eastern Highlands Province and a centre for processing and exporting coffee. However, it is informed by research carried out in Fiji, where Horst and I were able to interview senior officials associated with Digicel Fiji and Digicel Pacific who had extensive experience in PNG. Horst took the lead on research in Fiji, and has presented findings in a number of publications (see 2018, 2021; Sinanan et al. 2022) and a short documentary video on parenting in the age of the smartphone (see Horst et al. 2020). I took the lead on research in PNG, where I have investigated the subjects of nation making and globalisation since my doctoral dissertation fieldwork on mortuary ritual in 1984–85 (see Foster 1995, 2002, 2008). Research in PNG was conducted under the auspices of the PNG National Research Institute and in collaboration with the Department of Anthropology and Sociology at the University of Papua New Guinea (UPNG) in Port Moresby and the Centre for Social and Creative Media (CSCM) at the University of Goroka.[8]

At UPNG, Dr linus digim'Rina helped to recruit research assistants who interviewed friends, relatives and UPNG students about their mobile phone use and who also provided insights and observations about their personal experiences with mobile phones. At CSCM, Dr Verena Thomas facilitated research by making it possible for our project to sponsor a BA (Honours) student, Wendy Bai Magea, whose thesis research on the informal economy of airtime vendors was defined by the focus of the project (Bai Magea 2019). Bai Magea's thesis research was the basis for a short documentary film funded by our project and produced by Dr Thomas and other CSCM staff members, including the film's director, Dr Jackie Kauli (Thomas et al. 2018). Bai Magea and other research assistants also interviewed a range of Goroka area residents and university students about mobile phone use and contributed insights and observations about their own experiences with mobile phones.

Mixed Methods

Like all anthropologists, I appreciate the method of participant-observation. Participation in the case of this project involved first of all using a mobile phone in PNG with a Digicel or bmobile SIM card. This experience allowed me to learn firsthand the requirements for managing the pay-as-you go or prepaid subscriptions that most Papua New Guineans purchase in order to operate their phones.[9] These requirements included paying close attention to my account balance and developing a 'metered mindset' (Donner 2015) in addition to purchasing SIM cards at retail outlets and electronic 'top-up' (phone credit) and 'flex cards' (vouchers) from street vendors (Chapter 3). Once subscribed to the Digicel network, I received frequent text messages

---

8   Research in Fiji was coordinated through the School of Government, Development and International Affairs, Faculty of Business and Economics, University of the South Pacific in Suva. Our project sponsored the doctoral dissertation research of Lucas Watt (2019, 2020, 2023), which focused in part on the mobile phone practices of residents of an urban village in the greater Suva area.

9   'The vast majority of mobile customers in PNG have prepaid accounts and the market penetration of smartphones is growing (22 per cent in 2018)' (Highet et al. 2019: 21); 76 per cent of mobile connections in 2018 were prepaid (Highet et al. 2019: 19).

alerting me to promotions, giveaways and contests—key devices for making Digicel's market (Chapter 2). I also received late night calls from unknown numbers and, in one instance, a lewd text message, thereby introducing me to the sketchy world of 'phone friends' (Andersen 2013; Chapter 5). I eventually created a Facebook account that allowed me to monitor a range of pages that promoted discussion of current political doings (Chapter 5) as well as one page that specialised in voicing complaints about Digicel (Chapter 4). This account also allowed me to learn how my PNG Facebook friends addressed topics such as credit requests from family members and the dangers of fake online identities.

Observations made for this project were often informal and serendipitous. I was able to observe how people used mobile phones—or not—in public places, how they shopped for handsets in stores at Vision City (the main Port Moresby mall), and how store clerks and street vendors interacted with customers buying phone credit. These sorts of observations could be made almost anytime and anywhere in Port Moresby or Goroka. Other observations were organised around particular opportunities and events. I was able to accompany bank managers on their visits to check on stores that accepted mobile money and offered mobile banking services. These visits, one to a large settlement near the airport in Port Moresby and another to two trade stores in East Goroka, were extremely useful in educating me about the demanding on-the-ground logistics of digital financial services (Chapter 6). Such issues were taken up at a workshop that I attended held at the Grand Papua Hotel in Port Moresby and sponsored by the Pacific Financial Inclusion Programme, where I met individuals attempting to implement digital financial services in PNG whom I later interviewed.

Similarly, I was able to attend a program sponsored by the PNG Digicel Foundation in connection with its signature 'Men of Honour' campaign against violence, especially domestic violence. I also was able to make two separate visits to sites—one outside Goroka, the other outside Port Moresby—where the Digicel Foundation had constructed school buildings as part of its long-running education infrastructure initiative. These opportunities deepened my understanding of the role of corporate social responsibility in Digicel PNG's business operations (Chapter 6).

In addition to participant-observation, diary exercises and semi-structured interviews generated data about mobile phone use. Approximately 25 interviews were conducted by myself and research assistants in Port Moresby and Goroka with a range of users: old and young, male and female, rural and urban. These interviews often involved a recounting of all the mobile phones that a user previously owned and operated, including how the phones were acquired and, in many instances, lost or stolen. At UPNG, two dozen students kept detailed diaries of their phone use over a period of 48 hours. These diaries offered insights into how exchanges of small amounts of credit were a regular feature of romantic relationships (Chapter 3). Interviews with multiple street vendors in Goroka, three of which I conducted and the rest of which Bai Magea (2019) conducted, yielded insights into the habits of mobile phone users as well as information about the informal economy that grew around mobile phone repair and maintenance (Chapter 3).

Interviews with mobile users and vendors were supplemented and complemented by interviews with both corporate and government officials. Over the course of the project, I interviewed the then current CEOs of Digicel, bmobile-Vodafone and Telikom, as well as the Telikom CEO at the time of Digicel's launch in PNG. I was also able to interview other senior officials at Digicel PNG who were responsible for marketing, government relations and mobile money initiatives. In Suva, Horst and I talked with senior officials connected to Digicel Pacific and Digicel Fiji who were familiar with Digicel PNG's operations. In addition, I met with the heads of EMTV (PNG's lead public free-to-air television broadcaster) and a prominent Port Moresby advertising agency to discuss the implications of media convergence (Chapter 4).

I met several times with officials at the National Information and Communications Technology Authority of PNG (NICTA) to discuss topics that included SIM card registration, cybercrime policy and tower-sharing and universal access provisions. At the Independent Consumer and Competition Commission (ICCC), PNG's economic regulator and consumer watchdog, I met with officials to discuss consumer complaints about mobile communication services and rates. Meetings with officials at PNG DataCo Ltd, the state-owned operator of the National Transmission Network, were helpful in learning about plans for the country's telecommunications infrastructure. At the PNG Office of Censorship, I learned about the government's concerns over offensive materials disseminated through online platforms. All of these interviews and conversations with government and corporate officials were instrumental in developing an analytical approach that apprehends relations among consumers, companies and state agents as the unstable substance of a moral economy.

I drew on my prior experience studying transnational soft drink companies (Foster 2008) to make regular use of business news and corporate reports and press releases. This material offers more than just information. Read as much against as with the grain, these sources suggest insights into how corporations fashion their public image, project their attractiveness to investors and represent their moral obligations to stakeholders. Because Digicel Group Ltd is a privately held company, I was unable to buy shares and thus attend annual general meetings and receive annual financial reports. However, I was able to glean financial information about the company from a filing with the Securities and Exchange Commission submitted as part of an aborted initial public offering (Digicel Group Ltd 2015).

I similarly made use of articles about mobile communications and the information and communications technology (ICT) sector more generally that appeared in the two main PNG newspapers, the *Papua New Guinea Post-Courier* and *The National*. These articles were especially useful in narrating the controversies surrounding Digicel's licensing and launch in 2007 (Chapter 1), a story that has hitherto been told only in bits and pieces (see Barker n.d.; Duncan 2014; 'Ofa 2012; Watson 2011). Articles available online were compiled by Luke Gaspard, a research assistant at RMIT University, and a series of research assistants at the University of Rochester. By consulting extremely rare print copies of the *Post-Courier* and *National* held in the library of the PNG National Research Institute, I was able to view the striking advertisements that were part of Digicel's initial aggressive marketing campaign and Telikom's response to that campaign (Chapter 2). These advertisements document the massive subsidies on handsets and the deeply discounted promotional rates for voice calls and text messages that put mobile communications in reach of many Papua New Guineans for the first time. The ads also suggest how Digicel created a recognisable visual culture that for a time saturated the urban landscape of Port Moresby with red and white logos and brand imagery (cf. Willans et al. 2022).

In sum, just as this book attempts to decentre the ethnography of mobile use that is common in other anthropological accounts of mobile phones, it also attempts to take advantage of data drawn from a variety of primary textual sources. The book also benefits from the publications of other scholars who have given due attention to the uptake of mobile phones in PNG, especially the pioneering dissertation and extensive subsequent reporting of Amanda Watson (e.g. 2011, 2022; see also Lipset 2013, 2017, 2018).

Summary of the Book

This history of the mobile phone in PNG is partial in more than one sense of the word. It is partial first of all because it intends to be neither a comprehensive political and economic study of the telecommunications sector nor a thickly described ethnography of mobile phone users or employees of mobile phone companies. The book, moreover, pays more attention to Digicel than to its competitors, mainly because of the company's outsize role in quickly making and dominating a market for mobile phone use in PNG in the decade after liberalisation in 2005, during which rival Bemobile (later bmobile) underwent protracted rebranding and restructuring of ownership.

Furthermore, the book synthesises and analyses information and data gathered from a variety of sources and generated by participant-observation within the terms of a particular conceptual framework. The result is therefore a selective account of how mobile phones became a familiar part of social and material life in PNG, with varying effects for the consumers, companies and state agents linked to each other in a dynamic moral economy.

The result also reflects my own interests in the anthropology of both mobile phones and corporations. For example, I give close attention to the workings of prepaid subscriptions, the means by which mobile phone use has spread through the Global South (Kalba 2008). Digicel surely never claimed that its mission in PNG or elsewhere was to alleviate poverty, but its use of affordable pay-as-you-go vouchers in creating a market effectively employed the strategy of selling goods in small low-priced units that has been adopted by corporations attempting to reach poor consumers at the 'bottom of the pyramid' (Prahalad 2010; see Chapter 6). Attending to the everyday complications of managing prepaid subscriptions offers a perspective on mobile phone use different from that of many accounts based on research in the Global North, where users often operate on postpaid plans with unlimited talk, text and data.

Each chapter of the book performs a loose riff on a theme germane to the contemporary anthropology of mobile phones (e.g. infrastructure and appropriation) or the anthropology of corporations (e.g. market creation and corporate social responsibility), thereby making the details of the technology's history in PNG relevant to wider multidisciplinary discussions in media studies. Chapter 1 orients itself to the issue of infrastructure, a topic that attracted increasing attention across the social sciences and humanities as this research project unfolded (see Harvey et al. 2016; Horst 2013). I follow the lead of analysts who approach infrastructure with an eye towards its unruliness and contingency. This unruliness and contingency is a matter, so to speak, of both the liveliness of non-human things and the contested relations among human actors who seek to manage or govern the elements comprising the infrastructural assemblage on which mobile phone connectivity depends. In PNG, such an approach means paying equal attention to the physical challenges posed by an unforgiving topography of rugged mountains and dispersed islands and to the political challenges posed both by the strength of customary landowners to negotiate rents for cell towers and by the weakness of the national state to regulate the business operations of a foreign corporation and the agendas of more powerful international and multilateral actors.

In Chapter 2, Michel Callon's (2021) economic sociology provides a touchstone for briefly considering some of the ways in which Digicel made a market for its goods in PNG. Callon urges us to pay attention to the array of devices that orchestrate or frame encounters among goods, buyers and sellers—from matching algorithms to trade shows and window displays to advertisements and brochures. I focus on two devices: promotions and sponsorships. These devices not only capture attention by awakening curiosity and desire, but also facilitate attachments of buyers to goods by which things and persons become entangled or 'reciprocally constitutive of one another' (Callon 2021: 59). In the years following Digicel's arrival in PNG, such entanglements frequently materialised as string bags (*bilums*) in which people carried money, betel nut, mobile phones and other personal items. These string bags were netted with red and white fibres that formed Digicel slogans such as 'Expect More, Get More' or words of affection such as 'I Love Digicel – Best-Service'. Such affection is manifest in the letter that JY Bro sent to the *Post-Courier*, published 9 May 2008 under the heading 'Digicel "with the people"': 'Heartfelt thanks to Digicel for you have reached deeper into the most remotest [sic] areas of PNG and saved lives, including my mum' (quoted in Watson 2011: 49).

Chapter 3 considers the practice of appropriation—that is, of how users not only adopt but also transform new technologies, often putting these technologies in the service of unanticipated purposes and generating equally unanticipated consequences. But rather than simply celebrate such appropriation as evidence of human creativity, I follow François Bar et al. (2016) in regarding appropriation as one moment in a technology cycle that includes 'repossession', the appropriation of users' appropriations by the provider of the technology. The technology cycle connects users and providers—or consumers and companies—in an ongoing negotiation for control. Mobile phone users, for example, attempt to evade the constraints put on them by the corporate terms and conditions that apply to prepaid subscriptions; companies respond, in turn, to these evasive tactics with new constraints. Attending to these cycles of response and counter-response saves us from assuming that all market devices accomplish the aims of their designers and compels us to respect the insights offered by an ethnographic perspective on mobile phone use. It is precisely this ethnographic perspective that allows us to understand how in PNG the appropriation of mobile technology has been conditioned by prevailing conventions about the moral obligations associated with giving and receiving gifts.

Chapter 3 extends the discussion of making markets by looking at another device, the scratch-off voucher or 'flex card', and the market activities through which flex cards circulate. I argue that while indications of mobile phones stimulating entrepreneurial activity in PNG are faint at best, there is distinct evidence of the economic impact of mobile phones in the form of an informal economy centred on the buying and selling of prepaid credit and the maintenance of handsets. This informal economy, much like the various practices through which prepaid credit circulates, reveals how the different logics of commodity transaction and gift exchange become interwoven without one able to incorporate the other completely. The future of this informal economy is highly uncertain, however, as smartphones render the use of flex cards unnecessary by making online purchases of credit more convenient. Street vendors of flex cards and 'top-up', credit sent from the vendor's phone directly to the user's phone, understandably wonder if Digicel is repossessing the tools of their trade.

The spread of smartphones in PNG, especially in urban settings, meant that the mobile device was becoming in effect a hand-held computer. It also meant that telecommunications companies like Digicel were becoming something else—namely, internet service providers (ISPs) and entertainment and news companies. Chapter 4 examines these two different trends through the lens of media convergence, specifically, the economic and cultural dimensions of media convergence. Economic convergence resulted in Digicel's launch of a free-to-browse news service for network users and a prepaid digital television service that includes access to a free channel, TV WAN. Both the news service and TV WAN transformed Digicel into a content provider at the same time that the affordances of smartphones enabled users to become 'prosumers' or consumers who are also producers.

While prosumers could enjoy the freedom of creating their own customised online content—in the form of curated Facebook pages, for instance—they could never fully escape the constraint of having to pay a toll to their ISP in order to access the content they created. In other words, the conjuncture of economic and cultural media convergence underwrote a new business model in which not only social media companies (like Facebook) but also telecommunications companies (like Digicel) could generate revenue from prosumption, the latter by charging rent to network users for the opportunity to consume content that the users themselves had produced (see Foster 2007, 2011). This development pushed the power dynamics of appropriation—the struggle between consumers and companies for control over the technology cycle—into a new phase.

Chapter 5 treats the mobile phone as an instance of 'disruptive technology', but not in the primary business sense of Bower and Christensen's (1995) original formulation of the term (see Curwen et al. 2019). I instead consider how mobile phones entrained the disruption of a large variety of trust relations, relations that produce and are produced by the moral economy of mobile phones. These disruptions do, of course, have business implications. For example, trust relations between consumers and companies were upset by the advent of smartphones, as consumers questioned the accuracy and reliability with which companies like Digicel kept track of data usage (see Chapter 4). But disruptions entrained by mobile phones also affected trust relations between consumers and other consumers (that is, interpersonal relations) and trust relations between consumers (or consumer-citizens) and state agents. With regard to interpersonal relations, the kind of intimate and private one-to-one communication afforded by mobile phones disrupted social conventions about who could speak where, when and to whom. The impact of this disruption upset, in particular, trust relations between husbands and wives, as mobile phones in PNG like elsewhere came to symbolise the threat of extramarital affairs.

Disruption, however, was not confined to the domestic sphere. Smartphones, as the means of access to the internet, brought mistrust of state agents and agencies on the part of consumer-citizens into the public light of blogs and social media accounts. The affordances of smartphones made it easier to circulate and publicise criticisms of government policies and government officials, promising a new era of transparency and an end to official corruption. These criticisms in turn aroused the concern of their targets, state agents who responded with ICT regulations that some citizens interpreted as a constraint on their right to free speech and public expression.

Maintenance of trust and goodwill on the part of consumers is an important consideration for all business enterprises, and Digicel was no exception. In PNG, the company's main vehicle for generating trust and goodwill was the Digicel Foundation, a non-profit charitable organisation founded, according to the patron's message from Denis O'Brien:

> on the simple premise that wherever Digicel grows, our communities must grow with us; and this principle continues to fuel our ethos of giving back to our communities across the region. (O'Brien n.d.)

The Digicel Foundation is an exercise in corporate social responsibility (CSR), the topic of Chapter 6. My interest in CSR and the Digicel Foundation, specifically, is not to criticise the company's initiatives as somehow insincere or cynical—a kind of virtue signalling that doubles as covert marketing. On the contrary, Digicel's philanthropy delivered needed essential goods and services in the areas of health and education across PNG—goods and services of the kind that most people in the Global North expect as citizens to receive from the state. Accordingly, Digicel's CSR programs furnish useful examples for a discussion of 'post-political governance' (Garsten and Jacobsson 2007), a form of collaboration that promises to align the interests of corporations (and non-government and civil society organisations), on the one hand, and state agencies, on the other. Post-political governance blurs the boundaries between corporations and states, with the result that 'the people' make claims for basic goods and services in their capacity as consumers as much as in their capacity as citizens.

In a brief conclusion, I keep the question of citizenship in view by briefly returning to the matter of infrastructure with which the book begins. Infrastructural citizenship, a term and idea developed in the writings of urban geographer Charlotte Lemanski (e.g. 2018, 2022), refers to how everyday access to and use of infrastructure indicate and enact one's status as a citizen. It is often the case, for example, that people with restricted citizenship rights (immigrants and slum-dwellers) experience limited access to public infrastructure and sometimes use infrastructure as a tool of protest, as in blocking roads or bypassing metering devices. Infrastructure thus materialises the relationship between states and marginalised citizens in the mundane form of frequent power failures, gaping potholes and unclean water.

The entry of Digicel into PNG's liberalised mobile telecommunications sector was predicated upon the manifest incapacity of the state, in the form of the state-owned monopoly provider Telikom, to meet its responsibilities for service provision to all citizens. Digicel's immediate initial success in rolling out its network highlighted the state's incapacity and materialised a relationship between consumers and the company that—although the direct consequence of the state's action to liberalise the industry—promised to deliver what the state could not, namely, the semblance at least of inclusive infrastructural citizenship. But this promise has been challenged by subsequent developments in mobile communications, including the shift from telephony to data use, that have widened a digital divide between urban and rural areas and between more affluent and less affluent consumer-citizens. The future of infrastructural citizenship in PNG, specifically with regard to telecommunications services, is if not foreclosed then certainly uncertain.

The year 2022 marked a watershed moment in the history of telecommunications in PNG. Telstra's acquisition of Digicel Pacific was in the eyes of many observers motivated by geopolitical concerns about the growing influence of China in the region. The departure of Digicel, moreover, coincided with the arrival not only of Telstra, but also of Vodafone PNG, the subsidiary of a Fiji-based company that never surrendered its control of the Fijian mobile market to its upstart rival Digicel. This new moment raises new questions, ones that will likely preoccupy future research and writing on the mobile phone in PNG. Will Telstra and Vodafone PNG be able to narrow the digital divides in PNG and make good on the promise of infrastructural citizenship? How will the geopolitical agendas of Australia and its allies, including the US, on the one hand, and China, on the other, shape the mediascape of PNG and the Pacific region? In what directions will the moral economy of mobile phones continue to evolve, redefining relations among consumers, companies and state agents?

Part I.
After Liberalisation: An Evolving Mobile Market

1

The Politics of Mobile Phone Infrastructure: Licences, Towers and Gateways

Introduction: Stabilising the Network

Horst and Miller began their groundbreaking 2006 ethnography of mobile phones in Jamaica with a chapter on infrastructure. Echoing anthropologist Sidney Mintz's (1985) call to pay attention to the supply side of consumer goods, Horst and Miller asserted:

> Before we can appreciate the way low-income Jamaicans have integrated the cell phone into their lives, we first need to provide an understanding of how Jamaicans came to have these phones in the first place. (2006: 19)

Since then, a vibrant anthropology of infrastructure has taken shape. Recent studies, in addition to their renewed emphasis on the materiality of stuff and the liveliness of non-human things, urge us to think of infrastructure as more than pipes, cables and roads.[1] These studies, ethnographically rich and conceptually diverse, invite us to understand infrastructure as 'that assemblage of people, objects, practices and institutions on which both the realization and distribution of patterns of connectivity, movement, flow and presence are dependent' (Di Nunzio 2018: 2). For example, Ketterer Hobbis and Hobbis (2020) have described how the infrastructural assemblage that

---

1   A very small sample of the burgeoning literature includes: Larkin (2013); Starosielski (2015); Collier et al. (2016); Harvey et al. (2016); Anand et al. (2018).

connects rural Lau Islanders to the internet includes not only submarine cables and radio waves, but also human brokers based in the Solomon Islands capital of Honiara who circulate downloaded digital multimedia files as gifts in the form of MicroSDs.[2]

Similarly, the infrastructural turn in media and communication studies seeks to show how 'the material transport of information' in the form of audiovisual signals 'reframes traditional questions of media production, circulation, access, consumption, and policy and regulation' (Plantin and Punathambekar 2019: 165). Such an approach attends to 'the myriad ways people encounter, perceive and use media infrastructure' (Parks and Starosielski 2015: 7) and highlights the contingency and unruliness of infrastructures—features that seem manifestly relevant to apprehending not only the technical and logistical challenges that confront all mobile network operators in Papua New Guinea (PNG), but also the national and local-level political circumstances that conditioned Digicel's business plans (in particular) from the outset.

As Appel, Anand and Gupta (2018) persuasively argue, infrastructural assemblages furnish a stimulus and vantage point for looking anew at politics (see also Knox 2017a; Venkatesan et al. 2018). Their crucial role in distributing social goods that many people take for granted as the basis of everyday life—water, electricity, information—renders infrastructures peculiar instruments of governance. On the one hand, infrastructures by definition imply connectivity to an extensive network; that is, infrastructures recede into the distance—a long and winding road or a line of telephone poles disappearing beyond the horizon at the 'vanishing point' (Knox 2017b). At any moment, attentiveness to and curiosity about powers, so to speak, that lie beyond the horizon—a distinctive feature of Oceanic, especially Austronesian, cosmographies (Sahlins 2012)—can bring infrastructures into view as matters of concern and public inquiry (Latour and Weibel 2005). Such moments of 'infrastructural rupture' (Knox 2017a) generate questions that can compel political modes of engagement.

On the other hand, infrastructures extend into the private and intimate spaces of mundane existence—our kitchens and toilets and even, with mobile phones, our palms and pockets. This very immediacy and intimacy

---

2   Infrastructural assemblages can be conceived at different scales (cf. Strathern 1991). For example, the mobile handset is itself an assemblage of components such as rare minerals that travel along multiple supply chains with almost global reach (Mantz 2018). Operating the handset might require assembling a charged battery, a subscriber identity module (SIM card) and an airtime voucher (see Vokes 2018).

define infrastructures as sites of resistance and disruption, especially when they break down or become inaccessible. The capacity of infrastructures to deliver the prerequisites for everyday living is a measure of 'the morality and ethics of political leaders' (Appel et al. 2018: 22). Misuse or neglect of infrastructures as well as breakdowns can precipitate crises of political authority and the unmaking and redefinition of political subjects (Appel et al. 2018: 20).

Accordingly, I heed the suggestion to regard infrastructures as 'unfolding over many different moments' (Appel et al. 2018: 17)—as never fully completed processes. As Simone (2012) cautions:

> While infrastructure attempts to suture, articulate, or circumvent, its proficiency of engineering, substance of investment or institutional support does not guarantee that it will accomplish what it sets out to do.

This observation is particularly pertinent to many under-resourced Pacific Islands countries where failed infrastructure projects often materialise the state's broken promises to its citizens (Ketterer Hobbis 2018: 69; see Conclusion). A processual view of infrastructure thus highlights the uneasy and contested relations among various human and non-human elements within the infrastructural assemblage. I ask, specifically, what are the political effects of these relations? That is, 'how do infrastructures participate in and produce changing forms of the public and the private, of states and corporations, of citizens and consumers?' (Appel et al. 2015). How does the work of sorting out who owes what to whom—not to mention who is who—continually reshape the infrastructural assemblage (and thus the possibilities for communication) that enables and results from the proliferation of mobile phones? How has this work shaped the moral economy of mobile phones in PNG in the aftermath of liberalisation?

This chapter looks at mobile phone infrastructure in order to demonstrate how liberalisation quickly led in effect to a profitable private 'quasi-monopoly' on the part of Digicel in PNG (Network Strategies 2013). I discuss in turn the circumstances under which Digicel and state regulators attempted to stabilise three different elements of the infrastructural assemblage required to operate a mobile communications network. Each of these elements—licences, towers and gateways (submarine cables and satellites)—foregrounds a particular aspect of the moral economy of mobile phones, namely, the relations between companies and state agents. Each of these elements, moreover, generates its own peculiar friction and thus its own peculiar challenge for governing the infrastructural assemblage.

Licences

PNG is a founding member of the World Trade Organization (WTO), and the initial impetus to liberalise the telecommunications sector was part of the overall commitment to privatisation that WTO membership entailed. But the effort to liberalise mobile communications begun in 2005—leaving fixed line communications in the hands of the state-owned operator, Telikom PNG—met resistance from state actors including the incumbent mobile network operator, Pacific Mobile Communication Co Ltd (PMC), a Telikom subsidiary that began offering Global System for Mobile Communications (GSM) services in 2003. (PMC was also responsible for providing internet gateway services, leasing bandwidth from Telikom and reselling it to internet service providers under the brand name Tiare [Duncan 2014].) Digicel's gamble in continuing to build a mobile network infrastructure in the face of government orders to desist ultimately succeeded. The story of how the company kept its operating and spectrum licences is in part the story of how a foreign company was able to mobilise the support of international organisations to counter the objections of national politicians. It is also in part the story of how different views of and stakes in privatisation pitted different state actors against each other in a public media controversy that played out in the form of articles and letters published in PNG's two main newspapers, the *Papua New Guinea Post-Courier* and *The National.*

***

Pacific Islands nations were among the last countries in the world to liberalise their telecommunications sectors. While a history of the sector in PNG is beyond the scope of this chapter (see Sinclair 1984; Duncan 2014), it is important to note that in 1997, as a consequence of its WTO membership, PNG agreed to end the monopoly of the incumbent national operator in five years and to issue new operating licences within two years afterwards, by the end of 2004 (Barker n.d.). Under the government led by the late Sir Mekere Morauta from 1999 to 2002, the privatisation of Telikom was proposed. Amalgamated Telikom Holdings (ATH) PNG Ltd, a company jointly owned by ATH of Fiji and Datec PNG, was recognised as the successful bidder. Morauta's privatisation program, however, was ultimately put on hold by the new National Alliance Government led by Sir Michael Somare in 2002 (Barker n.d.), and the ATH PNG bid was rejected.

The Morauta Government established the Independent Consumer and Competition Commission (ICCC) in 2002. The ICCC was charged with facilitating competitive markets and safeguarding against monopolies. Barker (n.d.) notes that nevertheless the ICCC promptly extended Telikom's monopoly on both phone services and the international submarine cable gateway for an additional five years, until 17 October 2007. Meanwhile, the newly established Ministry of Public Enterprises, Information and Development Co-operation opened a second round of bidding for Telikom in 2004, in which Econet, a South Africa–headquartered company founded in Zimbabwe in 1993, was recognised as the winner. Econet's bid was also later rejected on the grounds that the company lacked the resources to develop PNG's telecommunications services (Barker n.d.).

Competition in telecommunications services commenced in the Pacific Islands region in 2002 in Tonga; Samoa followed in 2005, with Digicel starting operations there the following year. Despite its founding membership in the WTO, PNG telecommunications services remained a state monopoly until the National Executive Committee (NEC) announced in December 2005 that Telikom's exclusive control of mobile services would end in March 2006.[3] The ICCC was instructed to undertake public tender for two new mobile licences that would become operative in March 2007, seven months ahead of the date for ending Telikom's monopoly originally set by the ICCC in 2002. Although PMC had launched its upgraded GSM service in May 2003, marketed ever since variously and inconsistently as Bee Mobile, BMobile, B Mobile, BeMobile, Bemobile, bemobile, 'B' Mobile, B-Mobile or bmobile, it remained concentrated in a few urban areas and came at high costs to subscribers.[4] Barker (n.d.) reports that by mid-2007,

3   Barker (n.d.) recounts speculation at the time that this NEC decision was precipitated by discussions between PNG Prime Minister Sir Michael Somare and Prime Minister Thaksin Shinawatra of Thailand at the November 2005 Asia-Pacific Economic Cooperation (APEC) meeting in Busan, South Korea. Thaksin had made his fortune in the mobile phone industry, and speculators wondered if Thaksin and Somare would take an active interest in bidding for a mobile licence. The NEC noted in early 2006 that Thaicom/Shin Satellite, founded by Thaksin, 'had expressed interest in participating in the PNG telecom industry' (Post-Courier 2006a). When Thaksin reportedly visited Somare in PNG in 2008 after Thaksin's ouster from power, opposition leader Sir Mekere Morauta, in a critique of government corruption, commented: 'Did he [Thaksin] come all the way from exile in Dubai to pay homage to the Sepik River God [i.e. Somare] or to check out bemobile (phone company), LNG deals, timber deals, fish deals, construction deals, or other family businesses?' (PNG Attitude 2009).
4   Officially, BMobile was rebranded as Bemobile in November 2009, when 50 per cent of the company was sold by Telikom PNG (see ADB 2020: 1; see also below). However, one regularly encounters variations of the company's name in use. In this book, I attempt to follow the spelling that was current at any given time; bmobile was the name most commonly used in corporate marketing communications in 2022.

on the eve of Digicel's operations in PNG, Telikom supplied only 65,000 fixed lines while PMC had 160,000 subscribers; other estimates of mobile phone subscriptions at the time are even lower at 130,000–140,000 (Duncan 2014) and 100,000 (Cave 2012).

Many Papua New Guineans more than 10 years later bitterly remembered PMC's initial GMS service as effectively restricted to 'big shots' (see Martin 2013). The start-up package, which included a SIM card and 100 minutes of airtime, cost PGK125 (about USD44 in 2007), an amount far beyond the reach of most Papua New Guineans.[5] Mobile phones were, in those days, potent and portable symbols of wealth disparity and urban elites, not to mention plain evidence of an incompetent and uncaring state incapable of bringing development to the people and meeting its moral obligations to all citizens (see Watson 2011: 48–49).

Figure 1.1. bmobile-Vodafone headquarters, Port Moresby, 2017.
Source: Photo by R Foster.

---

5   100 toea equals 1 kina (PGK1.00). One kina was approximately equal to USD0.413 in 2014, USD0.305 in 2018 and USD0.284 in 2022.

Enter Digicel

The ICCC announced in January 2006 that tendering for two mobile network licences would open in March and close in May, with the winners to be revealed in October and formally licensed to begin operations in March 2007. The tendering process proceeded ahead of schedule and the awarding of two licences was made public on Friday 1 September. The two winning companies were Green Communications (GreenCom) and Digicel PNG, a subsidiary of Digicel Pacific Ltd, selected from 16 competing companies. ICCC Commissioner and CEO Thomas Abe noted that an important consideration in evaluating the tenders was the capacity of a company to roll out services in rural areas where 80 per cent of PNG's population resides (Post-Courier 2006b). Abe said:

> Mobile phone services will be available in many areas of PNG where at present, there is limited phone access and indeed it will bring mobile phones to a lot of places which currently have access to no telephone at all. (Fiji Times 2006)

The licence for GreenCom was granted to Dawamiba PNG Ltd, a partnership between the family of former Deputy Prime Minister Ted Diro and Dawamiba Engineering of Indonesia. A majority share in GreenCom was held by General Enterprise Management Services (GEMS), a Hong Kong–based private equity fund that listed Henry Kissinger as an adviser. GreenCom never began operations in PNG. The company 'collapsed under a reported PGK12m ($5.71m) debt in 2009' (Oxford Business Group 2012a), leaving local workers unpaid. In 2008, GEMS was part of a consortium that purchased 50 per cent of BMobile from Telikom for PGK130 million (USD61.87 million) in a sale forced by BMobile's poor performance in the face of Digicel's rapid expansion (for more on the controversial sale of BMobile see Barker n.d.; Watson 2011: 51ff).[6]

---

6   In 2013, an agreement for Vodafone Fiji to manage Bemobile collapsed; the agreement would have given the PNG Government's Independent Public Business Corporation (IPBC) a 51 per cent stake in the operator and a 40 per cent stake to the Fiji National Provident Fund. Subsequently, IPBC acquired 85 per cent of the shares of Bemobile Ltd, which operates in both PNG and Solomon Islands. In 2014, Bemobile signed a non-equity marketing partnership with Vodafone Group Plc, which lasted until 2019, and traded as bmobile-Vodafone. In 2017, the PNG Government announced the amalgamation of bmobile-Vodafone and Telikom along with DataCo, a state-owned entity created in 2014 to provide wholesale information and communication technology (ICT) services, under a new name, Kumul Telikom. In 2021, bmobile and Telikom merged as Telikom Ltd and Kumul Telikom Holdings Ltd was abolished.

Right from the start, reports surfaced in PNG newspapers that 'industry leaders' were concerned that the new schedule for ending Telikom's monopoly would be imprudent given that Telikom had not been able to make necessary upgrades to its network. These upgrades had been recommended by the Telikom board of directors in consultation with management firm KPMG (Post-Courier 2006a). In July 2006, PNG's minister for state enterprises, Arthur Somare (son of then Prime Minister Sir Michael Somare), announced that his government had dropped its plans to privatise a number of state-owned companies, including Telikom. In September, Telikom claimed that the conditions it required of the ICCC for surrendering its monopoly had not been met. Telikom initiated judicial proceedings through an Appeal Panel and took the ICCC to court in a bid to stop the issuing of licences ('Ofa 2011: 76).

Also in September, the government released a new National ICT (information and communications technology) Policy that reversed the NEC's 2005 decision to liberalise mobile services: the policy argued for staged competition and against the immediate issuance of licences to Digicel and GreenCom. The government acknowledged the pressing need to rehabilitate and upgrade Telikom's network capabilities at an estimated cost of between PGK500 million and PGK1.4 billion over five years ('Ofa 2011: 77). It also acknowledged that for technical reasons Telikom's network could not be interconnected with another carrier's network ('Ofa 2011: 77)—a problem that became apparent immediately upon Digicel's launch and that took several months to resolve (for details, see Watson 2011: 49–50).

In October, a month after the winning bidders were announced, the *Post-Courier* (2006c) reported that the Papua New Guinea Radio Communications and Telecommunications Technical Authority (PANGTEL) was still without a board of directors—a potential problem inasmuch as the board was the only authorised body to grant spectrum licences to any telecom company entering PNG. Amid this regulatory uncertainty, Digicel PNG nevertheless began to build the infrastructure for its network.

The new ICT policy unveiled in 2006 was never implemented, but it contained key recommendations that reappeared in another new National ICT Policy approved in June 2007, ICT Policy 2007. Most importantly, ICT Policy 2007 advocated a so-called NETCO/SERVCO model in which Telikom (as NETCO) would retain an indefinite monopoly over all telecommunications infrastructure (including the international submarine gateway) while licences would be issued only to service providers (as SERVCOs), and then only after June 2008 (see Post-Courier 2007a; Alphonse 2008). This policy, too, was never implemented, but it foreshadowed both the revocation two years later of the ICCC's authority to issue licences and the eventual establishment of PNG DataCo Limited, described on its website in 2022 as 'a state-owned entity, created in 2014 to provide wholesale services to the Information and Communication Industry, mandated to build, own and operate the National Transmission Network (NTN)' (PNG DataCo n.d.).

By early 2007, not only regulatory uncertainty but also the rift between Arthur Somare and Telikom CEO Peter Loko, on the one hand, and the ICCC's Thomas Abe, on the other, had become public. Abe for his part kept the ICCC's process on track, issuing carrier licences on 27 March and declaring operations to begin on a date after 1 May but before 17 October. Loko, in the wake of the National Court's rejection of Telikom's application for an interim injunction on the licences issued by the ICCC, alleged that the new mobile companies would invest their profits outside PNG (Post-Courier 2007j). Loko thought that PANGTEL should be making licensing decisions, not the ICCC (personal communication, 1 August 2015). Abe responded with the suggestion that Telikom accept both competition and the fact that it had been an inefficient monopoly.

The dispute over licences and the future of competition in mobile services reached a climax when the government approved ICT Policy 2007 on 21 June. The NEC announced that the introduction of new carriers would be pushed back by another year and the licences previously issued to Digicel and GreenCom would be revoked. In early July, the National Court declared that the issuance of the two licences was proper and lawful, and the ICCC accordingly refused to revoke them. PANGTEL had issued Digicel a spectrum licence on 20 June (Post-Courier 2007b), and Digicel planned to launch its network in July (Post-Courier 2007c). But PANGTEL Acting Director General Charles Punaha reported that the government had requested PANGTEL to review the newly amended policy in order to verify that there were no impediments to the policy being implemented (Post-Courier 2007b). Punaha's statement implied that PANGTEL was a statutory agency that made decisions independently of Minister Arthur Somare— a point subsequently reiterated in the face of charges made by Digicel that PANGTEL's actions were 'politically motivated' (National 2007a).

Predictably, the business community in PNG responded unhappily to the announcement that the introduction of new mobile services was being deferred and worried about the effects of the government's backflip on foreign investment. Letters to PNG's two daily newspapers similarly expressed frustration with Telikom's services and PANGTEL's actions. One letter writer asked: 'Where is the logic in denying consumers their rights and competition?' (National 2007b). Government officials, in turn, invoked other concerns. Sumasi Singin, chairman of the Independent Public Business Corporation (IPBC), Telikom's sole shareholder, told the *Post-Courier* that the government sought to preserve national security by preventing other parties from operating separate telecommunications networks. Singin explained, moreover, that the newly amended ICT policy would allow competition in fixed line and internet services as well as mobile services. Both proponents and critics of ICT Policy 2007 thus claimed on different grounds—sovereignty versus development—to represent the best interests of the nation (Post-Courier 2007d).

Digicel's Response

Throughout the period of turmoil surrounding the validity of its licence to operate, Digicel pursued several strategies in order to commence operations on schedule, all the time working to establish the infrastructure for its network. (Peter Loko would, in retrospect, contrast the way in which Digicel took the mission-driven risk to plunge ahead with the way in which Telikom had to wait for approval from IPBC before acting; personal communication, 1 August 2015.) Digicel director Seamus Lynch reported meeting with Prime Minister Sir Michael Somare in early March 2007 and receiving assurances that Digicel's licence would be protected (National 2007a).[7] In May, an arbitrator was brought to PNG from Australia at the behest of the ICCC in order to hear Telikom's grievances, which Digicel's lawyers rebutted (Post-Courier 2007e). And in the weeks leading up to Digicel's anticipated launch in mid-July, CEO Vanessa Slowey lobbied the PNG Business Council, emphasising how Digicel would provide not only superior service to Papua New Guinean customers, but also investment in the country and direct and indirect employment of thousands of citizens (Post-Courier 2007f).

---

7   The Prime Minister later publicly defended Arthur Somare's decision and asserted that Digicel was operating outside the country's policies (Post-Courier 2007g).

Digicel officially opened for business in Port Moresby and Lae, PNG's two major cities, on Friday 20 July 2007. In a letter dated the same day, Charles Punaha informed the company that PANGTEL, acting in accordance with the newly amended ICT policy, had determined Digicel's spectrum licence to be null and void. PANGTEL demanded that Digicel cease operations and halt all work on its network infrastructure. PANGTEL also returned to Digicel two cheques for licensing fees amounting to PGK918,437.42 (approximately USD333,000) (Post-Courier 2007h). PANGTEL's letter reached Digicel on Tuesday 24 July, by which time some 20,000 Papua New Guineans had already signed up for Digicel's mobile services. The company immediately sought and won a court injunction against PANGTEL. Seamus Lynch proclaimed 'Any rumour that Digicel is going out of business is rubbish' (National 2007a). Digicel celebrated with a festive launch party that evening at the glitzy Lamana Gold Club in Port Moresby.

The scene was now set for a protracted struggle between Digicel and the ICCC on one side, and Telikom, PANGTEL and Arthur Somare's Ministry of Public Enterprises, Information and Development Co-operation on the other side. Newspapers published duelling statements from the ICCC and PANGTEL in which each agency claimed that it and not the other had duly followed relevant national laws and procedures (see Abe 2007; Punaha 2007). The PNG state was hardly acting monolithically in this case, as previous court decisions against Telikom attest. Digicel, moreover, received support from not only the ICCC but also political and business interests committed to liberalisation sooner rather than later. Digicel would ultimately win this struggle, although ICT policy would be revised in 2009 to strip the ICCC of its authority to issue carrier licences—a prerogative reserved for a new agency, the National Information and Communications Technology Authority (NICTA), that would replace PANGTEL.[8] In October 2007, a new ministry for information and communication services was created by Prime Minister Sir Michael Somare's government, to which Arthur Somare yielded his responsibility for PANGTEL.

***

8   In 2014, NICTA awarded an operator licence to Dubai-based Awal Telecommunications Corporation Limited, which expressed interest in providing mobile services. Four years later, the company had not set up business operations and the *Post-Courier* reported that it had disappeared 'without a trace' (Post-Courier 2018a).

The moral economy perspective taken in this book directs attention to how Digicel was able to leverage resources from outside PNG in its struggle with agents and agencies of the PNG state. A few days after PANGTEL revoked Digicel's spectrum licence, *The National* (31 July 2007c) ran an article misleadingly titled 'World Bank Backs Digicel'. The article quotes from an International Finance Corporation (IFC) document disclosing the investment of up to USD40 million in the form of a loan to Digicel Pacific Ltd, the majority shareholder of which was Denis O'Brien, in support of the construction of a nationwide GSM network.[9] The article, which oddly does not mention the loan, also includes comments from Digicel CEO Vanessa Slowey, who said 'The people of Papua New Guinea have been deprived of an affordable and reliable mobile telecommunications service for too long'. Slowey added 'we are not going away and Telikom and Pangtel need to know that Digicel is here to stay'.

Digicel's access to powerful multilateral organisations was also indicated by a report that the head of the European Union (EU) delegation to PNG had expressed concern over the government's handling of the dispute between Digicel and Telikom (Post-Courier 2007i). Similarly, the company's capacity to enlist support from foreign governments was demonstrated by the visit to Port Moresby of Ireland's Ambassador to Australia and PNG, Mairtin O'Fainin, who met with PNG government officials to discuss the investments of prominent Irishman Denis O'Brien's company (Post-Courier 2007k). Less visibly, the threat of discipline from the WTO, whose regulations on competition PNG was obliged to follow, might have played a role in tempering moves to prevent Digicel from building and operating its own mobile network ('Ofa 2011).

Digicel was thus able to evade restraints on its business operations by exploiting disagreements between two PNG state agencies, PANGTEL and the ICCC, and by mobilising support from multilateral organisations (the EU) and foreign governments (Ireland). Fundamentally, however, the moral legitimacy of Digicel's claim to continue operating in PNG rested on its success in delivering long sought-after services to PNG consumers. The company surely gambled in setting up its network infrastructure months before its carrier and spectrum licences had been approved and by launching operations while a major court case against the company was still pending. This same strategy backfired in Solomon Islands, where Digicel lost its bid,

---

9   See: disclosures.ifc.org/project-detail/SPI/26295/digicel-png, accessed 26 January 2021.

ironically, to Telikom's subsidiary, BeMobile (Solomon Times 2008). The observation of Ed Willett, a former official of the Australian Consumer and Competition Commission, is apposite. Willett (2008: 105) spoke at a January 2008 workshop on ICT policy in Port Moresby:

> The saving grace for Digicel and other prospective service providers may well lie in the fact that liberalisation of markets is usually politically difficult. Once consumers have tasted the benefits of competition, they are generally far more attuned to the issues.

Once Digicel had enrolled its first batch of subscribers, reversion to the status quo ante was practically impossible. Peter Loko similarly noted that once the majority rural population of PNG—to which Digicel had paid far more attention than Telikom—had gotten a taste of mobile services, there was no way that anyone was going to take these services away (personal communication, 1 August 2015).

Towers

Figure 1.2. Cover image of a GSMA report.
Source: www.gsma.com/mobilefordevelopment/blog/papua-new-guinea-how-can-mobile -technology-be-harnessed-for-digital-transformation, accessed 6 November 2023 (Highet et al. 2019).

The image of a solitary cell tower perched atop a high mountain amid lush, rugged terrain signifies the perils and promises of delivering telecommunications services in places like PNG. Digicel officials frequently referred to the challenges of installing towers, which served as icons of the company's resilience and its commitment to extend services to remote rural populations. Denis O'Brien once told a *Washington Post* reporter:

> Think of building a network in a country the size of France, with hundreds of islands, no roads and cerebral malaria … I've lost nine colleagues in Papua New Guinea building our network—from plane crashes, accidents, car crashes … It is the toughest of tough places. (Metcalf 2013)

Stabilising the network—assembling the infrastructure and keeping the assemblage intact and functioning—requires constant effort.

PNG has no provisions for tower sharing. When Digicel entered the country in 2006, it immediately began to construct its own network of towers, rolling out the first 120 at a cost of USD160 million and moving into places that Telikom and its subsidiary BMobile had never gone. By October 2020, Digicel PNG CEO Colin Stone could claim that his company had invested over USD1 billion in 'connecting the unconnected', providing coverage for more than 85 per cent of the nation and promising to build towers in 45 new places by March 2021.

> 'That's our main focus,' Stone said, 'to give rural people the same benefits that people here in Port Moresby have. If Digicel hadn't done this, then the people would still be unconnected'. (National 2020a)

Locating and maintaining cell towers, the great majority of which run off solar panels with backup diesel generators, involves lots of work. The story of how Digicel expanded and developed its network of towers turns on how the company successfully engaged in private–public partnerships, on the one hand, and dealt with the demands of disgruntled landowners, on the other. Since its initial rollout, many but not all towers have been upgraded from 2G to 3G and 4G LTE. The result has widened a digital divide between urban and rural areas that has deferred the promise of national unity and equal access to telecommunications services that accompanied the liberalisation of mobile services in PNG (see Conclusion).

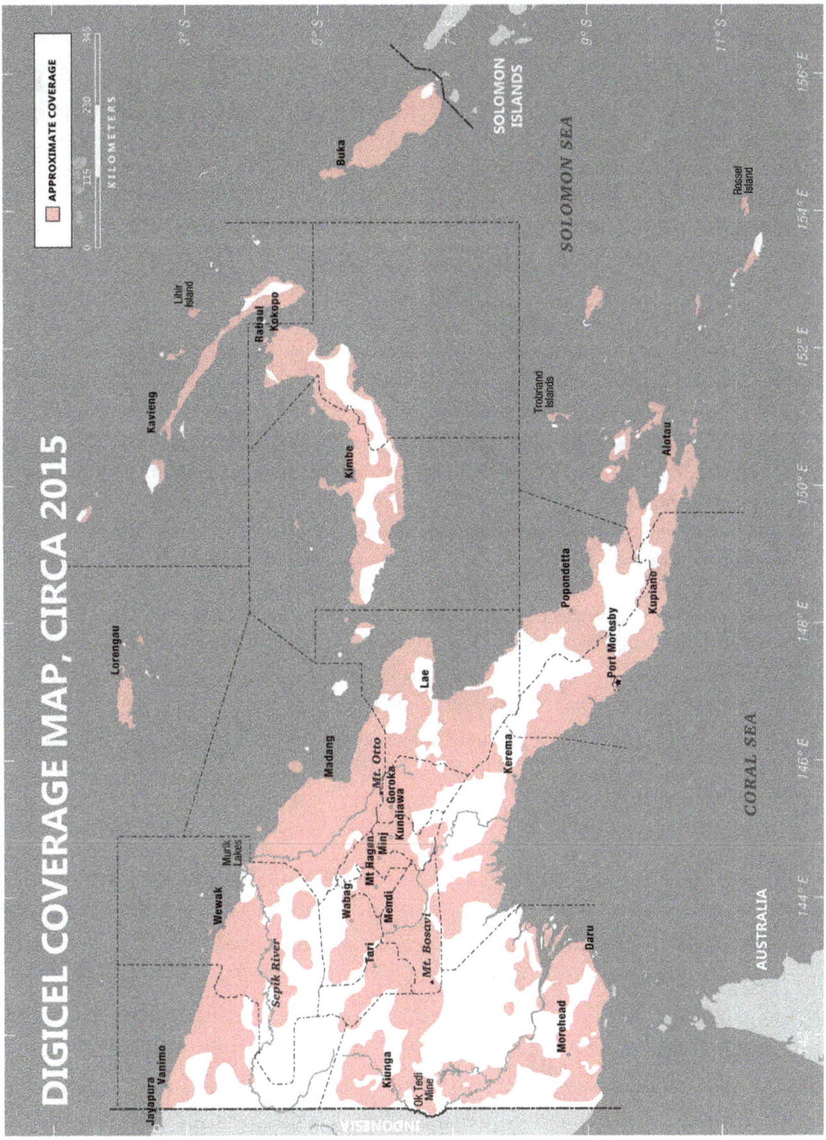

Map 1.1. Digicel coverage map, circa 2015.
Source: The Map Shop.

> **Box 1.1**
>
> *Voices of the Rainforest* (Feld et al. 2019) is a concert film that immerses audiences in the sounds and images of a day in the life of the Bosavi rainforest region (Papuan Plateau) and its human and non-human inhabitants. These inhabitants include the Kaluli people, who dramatically enact their paramount cultural value of reciprocity in the moving ceremony known as Gisaro—the focus of Edward L Schieffelin's (1976) evocatively titled ethnography *The Sorrow of the Lonely and the Burning of the Dancers.*
>
> At the end of the film, Monika Degelo offers a poignant community statement, 'delivered using a parallelistic oration structure more reminiscent of earlier generations of Bosavi men' (S Feld, personal communication, 26 April 2023). The statement, unscripted and unprovoked, was filmed in Bona village, Bosavi, in June 2018. It is, among other things, a remarkable formulation of the moral economy of information and communication technologies.
>
> Monika Degelo begins by noting that in the past, birds were messengers who told people when rain or a death or a visitor or a witch was coming. Then the missionaries arrived with the first radio, and instructed the people to stop listening to the voices of the birds. Like the birds, the radio sent messages about the comings and goings of sick people and planes. But when the battery wore out, the mission refused a request for a new one. The people asked the government for a radio to arrange emergency air transport, but the government did not give one. So, the people asked companies, like Oil Search, for a radio, but the companies did not give one or feel sorry for the people. Now the people cannot 'hear from the outside' and are unable to know if a plane will arrive when they transport a sick person to the airstrip.
>
> 'What can we do?' Monika Degelo concludes with an appeal to the anthropologist and filmmaker Steven Feld and his American friends to help put a Digicel cell tower in Bona. 'That's what I would like.'

In 2014, Digicel began an expansion of its network with Project Discovery, an addition of some 300 towers to the network, many located in out-of-the-way parts of the country with sparse populations.[10] To document the project, the company sponsored a video, *Digicel's Project Discovery PNG 2015*, that ran on TV WAN, Digicel's television service offered on the company's Digicel Play platform (see Chapter 4). The video highlights the physical challenges of erecting towers on desolate mountaintops where crews must camp for weeks while tons of materials are brought to the site by helicopter. Digicel's logistical and technical prowess, from satellite mapping of appropriate locations for towers to hooking up the generators that power towers in the absence of an electrical grid, are front and centre. But the narrative mentions other elements of the infrastructural assemblage that must be put in place in order for mobile phones to work. I concentrate here

---

10   An image of Digicel's coverage map circa 2015 can be found in Jorgensen (2018).

on two of these elements, each of which introduces its own kind of friction into the never-ending process of holding the infrastructural assemblage together: landowner agreements and access to capital and credit.

The Sorrow of the Landowners and the Burning of the Towers

In 2015, then Digicel CEO John Mangos claimed that 95 per cent of the company's towers were operating without problems, while 5 per cent were not. Mangos pointed out, however, that although there were a few regular 'hotspots', the 5 per cent were constantly changing (personal communication, 19 March 2015). Two years later, Oliver Coughlan, one of Mangos's successors as CEO, lamented the ongoing problem of vandalism for Digicel's network. He noted the 'costly exercise' involved in restoring vandalised towers:

> Many of the areas these towers are erected in a [sic] very remote places and time is lost trying to move not just parts but people, most times by helicopter, into these areas to fix them.

Coughlan admitted:

> We don't understand why people would want to destroy these facilities which in turn could be detriment [sic] to somebody who may be suffering from a health issue and may need help. (Post-Courier 2017a)

The problem of vandalism arose almost immediately after Digicel launched its network in PNG and separately from the problem of stolen solar panels and diesel fuel that power the majority of towers. In 2010, only 160 of the 700 towers maintained by Digicel nationwide were linked to PNG Power; only 12.4 per cent of households in PNG had access to electricity (Oxford Business Group 2012a). In 2018, only 300 of Digicel's 1,100 towers were on the grid (James 2018).

From the very beginning, one of the causes, if not the primary cause, of vandalism has been disgruntled landowners (see Box 1.2). Because many of Digicel's rural towers are located on land held by customary ownership, which constitutes about 95 per cent of all the land in the country, the company deployed a large staff to manage the approximately 1,100 separate leases for the towers in its network. Digicel CEO John Mangos said that the negotiations with landowners were continual. He observed that towers were not like mines or other development projects tied to one particular piece of land; if the company were required to remove a tower and put it somewhere else, that would be acceptable (personal communication, 19 March 2015).

Box 1.2

Letter to the Editor, *Post-Courier*, Monday, 6 July 2015

RESTORE SERVICE

With the introduction of Digicel into the telecommunication industry in 2007, communicating through mobile phones has increasingly become entrenched in the lives of Papua New Guineans. More people now have access to mobile phones. Communicating through them is fast, convenient and cost effective. Recently, I have made numerous calls to my friends and families at home in the Anglimp area, but I have not been able to get through. Almost all my calls have failed. I have been told that this is largely due to the closure of the Digicel tower at the Kuli Gap Mountain, which serves mainly the Anglimp area, among others. The tower was closed after some issues arose between the landowners and Digicel over the operation of the tower. It has been a while now since the tower was closed, but no remedial action has been undertaken by anyone. Innocent phone users in the Anglimp area, as well as those who have contacts for people in the Anglimp area, have been denied access to reliable communication service. We want to have access to reliable service to communicate with our friends and loved ones anywhere and anytime, as currently enjoyed by the people in the Anglimp-South Waghi electorate and elsewhere. Who is responsible for restoring this vital service, as far as accessibility of reliable communication service for the affected people is concerned? As custodians of the people, our responsible leaders and agencies in the Anglimp area, including our MP [member of parliament], should look into this issue and find ways through which they can quickly resolve it. Try to settle the outstanding issues between landowners and Digicel, and allow Digicel to operate the tower, or if not, find another area where the tower can be put to restore reliable communication service for the people of Anglimp. A reliable communication system is vital for individuals, the operation of schools, health centres and other agencies that are there in the Anglimp area.

Rablam

The Project Discovery video succinctly describes the process by which landowner agreements are ideally arranged. Landowners are identified with the assistance of the PNG Department of Lands and Physical Planning and local leaders, especially in cases when there are conflicting claims to ownership. A memorandum of agreement (MOA) is then signed for use of land and rental payments. Nancy Sullivan reported in 2010 that annual rental payments to landowners in East Sepik Province amounted to PGK3,000 and a contemporaneous newspaper article reported the demand of these landowners that the monthly land lease rent be increased from PGK260 per month to PGK5,000 (National 2010a). Rents paid for towers in urban areas are said to be higher.

The video points out that Digicel staff maintain relations with landowners during and after the construction phase of a tower. The maintenance and security of towers provide local employment. Security for the towers is handled by four 'guardians', each of whom is given a phone to report problems, including 'theft of equipment' and 'any violence associated with the tower'. The video recognises that these problems usually stem from 'landowner issues' and 'rental payments', which Digicel negotiators attempt to resolve. Recurrent vandalism, along with challenges involved in supplying power to the towers, are major causes of network destabilisation.

The disgruntlement of landowners arises from two not mutually exclusive circumstances. First, there are disputes regarding who is in fact the rightful customary owner. In 2008, members of three clans jointly attacked Digicel employees and destroyed diesel generator sets at the Mt Otto tower site in the eastern highlands. The clan members contended that a fourth clan, Sehayuha clan, 'conspired with Digicel to erect the tower on customary land in dispute' (Pacific Islands Report 2008). In 2011, 50 members of Sehayuha clan converged on the Mt Otto site, where Telikom and Bemobile also had towers, and 'switched off all the facilities': 'Landowners claimed that Digicel PNG had been operating on their customary land setting up a huge tower without settling their demands for land compensation' (National 2011). A few days later, Digicel's security operation manager presented chairman of Sehayuha clan Veho Mohokule with a cheque for PGK58,000 and the matter was resolved.

In 2014, as part of Project Discovery, Digicel undertook construction of a tower in the Karawari river valley at a site that two different groups claimed as their customary land. Although the tower was completed, tensions simmered until 2018, when members of both groups vandalised the tower, rendering it inoperable. Leaders from both groups subsequently met to seek a solution, managing to recover most of the stolen solar panels. In this instance, however, Digicel reportedly declined to repair the tower, 'deeming it too expensive for the small amount of traffic it received' (Newens 2021).[11]

Second, disgruntlement arises over the amount of rent that Digicel pays to customary owners as well as the promptness with which such payments are made. Sullivan (2010b: 11) noted that East Sepik landowners felt misled

---

11 In April 2023, Timothy Andambo, the head of a landowner's association in Enga Province, attributed violence that led to the deaths of six people to a dispute over a Digicel network grid tower (Radio New Zealand 2023).

by the MOA they signed, understanding the unfamiliar term '3000 kina per annum' to mean a monthly payment of 3,000 kina. A blog post by a self-identified 'former Digicel employee' alleges that Digicel routinely signs rental agreements with willing but illegitimate parties claiming to own customary land (PNG Attitude, 23 December 2018, Digicel Needs to Come Clean with Illiterate Landowners). The company also allegedly made partial rental payments, compelling aggrieved landowners to come to Port Moresby to inquire about their missing rent.

Such situations of course look different from the company's perspective. A vandalised tower in Central Province was attributed to the complaint that the principal customary landowner had not been paid for three years. Digicel CEO John Mangos explained that the original agreement for the tower had been signed with a group that was subsequently deemed illegitimate by a local landowner court:

> We have attempted in good faith to negotiate for the use of the land. However, during negotiations with the new landowners, the tower has been continually vandalized with the cable cut and equipment stolen. (National 2010b)

Mangos said that the company was ready to remove all its equipment and explore legal options with regard to its losses.

Land disputes introduce considerable friction into the process of resource extraction throughout PNG, and they similarly condition the shape and size of the country's networks of cell towers. The process of building and maintaining such a network involves more than arduous efforts to transport materials by barge or helicopter. In PNG, landowners—human actors in the infrastructural assemblage—often prove more difficult to control than non-human elements such as unforgiving terrain or extreme weather.

Funding the Network

Disgruntled landowners exercised their capacity to shrink and destabilise Digicel's network. But the company demonstrated its own capacity to expand and strengthen its network and thus extend its dominant market position, for instance, by gaining access to financial resources through various public–private partnerships. These partnerships provided Digicel with capital for infrastructure projects and goodwill with both multilateral institutions committed to competitive markets and local populations desirous for reliable telecommunications services.

Digicel's entry into the Pacific region quickly raised concerns among the cash-strapped state-owned telecoms confronted with a new and nimble competitor. In PNG, Digicel committed USD160 million to building its network, of which USD40 million was contributed through a low-cost loan from the International Finance Corporation (IFC), a member of the World Bank group.[12] According to its website, the IFC 'advances economic development and improves the lives of people by encouraging the growth of the private sector in developing countries' (International Finance Corporation n.d.). IFC invests in companies, financial institutions, and other businesses that are majority-owned by the private sector and that operate in IFC's developing member countries.

Some officials at incumbent national telecom operators perceived a conflict of interest in IFC loans. In 2008, Marty Robinson, CEO of Our Telikom in Solomon Islands, declared:

> Our Telikom considers it inappropriate for the World Bank to be going around the Pacific Islands promoting competition in the telecommunications sector and at the same time providing finance to Digicel through IFC to perhaps influence them to enter markets that it would otherwise not consider. (Tabureguci 2010)

Major expansions and upgrades of Digicel's network of towers in PNG have been made possible through access to finance from development organisations. The IFC provided further loans of USD80 million in 2009 and USD26.8 million in 2011. The Asian Development Bank (ADB) offered a USD25 million loan in 2009 to launch the ADB Digicel Mobile Telecommunication Expansion Project, which expanded Digicel's coverage into PNG's islands region, including Manus, Alotau and the Trobriand Islands. Vanessa Slowey, then Digicel Pacific CEO, commented:

> Communication is a basic human right, and ADB helped us make this right a reality for the people of Papua New Guinea, many of whom never had access to communication services. (ADB 2012)

ADB also funded the expansion of Bemobile in Solomon Islands in 2011 with a loan of USD40 million, making an equity investment of USD9 million in the company as well.

A similar project also commenced in 2011 with the goal of bringing communication services to the sparse and widely distributed population of Western Province. Costing USD26 million, the project was sponsored by the

---

12   Export Development Canada contributed another USD25 million.

PNG Sustainable Development Program (PNGSDP), a fund established to compensate the residents of Western Province for ecological damage caused by the Ok Tedi mine. Digicel won the tender to install about 50 towers across the province. While these towers were opened to use by other service providers, Digicel was well positioned to use the opportunity to enlarge its existing network. In 2018, PNGSDP allocated USD32.5 million for Digicel to upgrade the company's 78 towers in Western Province for 4G services and to build an additional 19 new towers to be owned by PNGSDP (Business Wire 2018).[13]

Digicel's Project Discovery, which added 300 towers to the network in remote locations in the Sepik River area and highlands region, also received IFC support. IFC offered approximately USD1.7 million in trial support of establishing solar-charging stations for mobile phones in rural villages with no access to electricity. IFC saw the charging stations as an opportunity to develop a network of 'solar entrepreneurs' and a market for off-grid lighting solutions (International Finance Corporation 2013). Digicel's vested interest in helping rural consumers overcome barriers to purchasing and using the company's services should be self-evident.

Public–private partnerships are often touted as win-win arrangements (see Chapter 6). Public good and private interest are satisfied at the same time; the market delivers development to the people and profits to businesses. In PNG, the same not-so-invisible hand also generates political capital for members of parliament (MP) who choose to make a deal through Service Improvement Program funds assigned by the national government to their districts. Newspaper reports periodically announce how an MP has contracted with Digicel to erect a tower in his (always his) district. The contract stipulates that Digicel covers half of the PGK1 million cost while the local MP covers the other half. Publicity photos often feature the handing over of an oversized mock cheque from the MP to a Digicel representative. The exchange enables the MP to redeem his campaign promise to bring development to the district—from Telefomin to Tagula—while at the same time subsidising the expansion of Digicel's network.[14]

It might seem peevish to criticise partnerships that make telecommunications services available to remote rural populations. Digicel accomplished this feat. But it is not unfair or irrelevant to point out that these populations

---

13 Money was also allocated to the Digicel Foundation for building schools (see Chapter 6).

14 The late Peter Loko, CEO of Telikom when Digicel began operating in PNG, claimed that Digicel copied the partnership model for erecting towers from Telikom—one of the only things they copied, he added (personal communication, 1 August 2015).

must pay to use Digicel's services. Nor is it beside the point to observe that these partnerships, like the financial assistance received from IFC and ADB, subsidise the infrastructure of a privately owned for-profit company whose entry into PNG was justified in terms of the superior benefits of market competition when compared with state-managed services.

Widening Digital Divides?

Creating its own extensive and exclusive network of towers—many of them located in places where Telikom had never built towers previously—gave Digicel an enormous advantage over any future competitor, including the incumbent BMobile, who would have to construct towers of its own.[15] It also entailed the problem of interconnectivity that was not resolved until months after Digicel launched. Yet while interconnectivity was a problem for Digicel consumers who could not call Telikom fixed lines (at government offices, for example) or BMobile customers, it is unclear whether it was a problem for Digicel. Peter Loko, CEO of Telikom during Digicel's July 2007 launch, claimed that Digicel never asked Telikom to negotiate interconnectivity despite knowing full well what this would mean for its customers (National 2007d). Digicel director Seamus Lynch reportedly claimed that interconnectivity would not affect customers as more and more subscribers signed on (National 2007a). Put differently, as more and more customers signed on to Digicel, the burden imposed by lack of interconnectivity would shift on to the shoulders of Telikom and BMobile service users.

Owning and operating the network, however, came at a cost to Digicel in addition to the costs incurred by vandalism and theft. In order to keep up with the transition from basic phones to smartphones and the accompanying increase in demand for data, Digicel had to upgrade its 2G towers to 3G, 4G and LTE. These upgrades were done selectively. John Mangos reported in 2015 that upgrades from 2G to 3G were done on a tower by tower basis (personal communication, 19 March 2015). Demand for data, of course, is most intense in urban areas, where upgrades to 4G LTE were first rolled out in Port Moresby in 2014 for postpaid customers. According to a 2019 GSMA report:

---

15 The issue of third-party tower sharing was still not resolved in 2022. Vodafone PNG, which began operations in 2022, decided to construct its own network of towers after discussions with both Digicel and Telikom/bmobile about tower leasing and sharing failed to produce a workable arrangement (Post-Courier 2022a).

> Internet usage is skewed towards urban centres, with almost 70 per cent of internet users residing in the cities of Port Moresby and Lae … Internet users tend to be young—almost half are aged between 18 and 24 (48 per cent) and 82 per cent are under the age of 34. (Highet et al. 2019: 24)

Likewise, access to electricity—handy for recharging power-hungry smartphones—is greater in urban areas. Murdock (2022) notes that between 10 and 15 per cent of PNG's population has access to on-grid electricity, although 60 per cent of the population 'is connected if off-grid solar products are considered'.

In 2018, 2G services were still common in rural areas, where 87 per cent of PNG's estimated 8.42 million people were living; 2G connections accounted for 55.5 per cent of the market in PNG, with 3G and 4G connections each accounting for about 25 per cent (Oxford Business Group 2019). Digicel reportedly aimed to cover 80 per cent of PNG's population with 3G by 2020 as well as to upgrade 173 towers from 3G to 4G. bmobile, which has competed respectably with Digicel in urban areas, aimed to command 20 per cent of the 4G market by 2019, a goal made more possible by absorbing Telikom's 4G services and some 400 base stations (Oxford Business Group 2019), in which Telikom had invested when migrating its short-lived and unsuccessful CDMA Citifon service (2011–16).[16] (Telikom exited the mobile market in 2019, leaving Digicel and bmobile as the only two competitors.)

It is unclear if Digicel and bmobile achieved their specific goals, given the onset of the global COVID-19 pandemic in 2020. Telikom/bmobile did, however, expand access to 4G, announcing at the end of 2020 upgrades that allowed REDiSIM (3G) users to access the company's 4G network without having to change SIM cards (Telecompaper 2021). At the end of 2022, Telikom/bmobile required REDiSIM users whose numbers begin with 75 or 76 to acquire an upgraded 4G SIM, which the company provided free of charge and with a bonus of 10 gigabytes (GB) of data and unlimited On Net calls (National 2023). A GSMA (2023) report, moreover, indicates that 2G connections in PNG, with a subscriber penetration rate of 41 per cent in 2022, had declined to 26 per cent of the market while 3G connections increased to 31 per cent and 4G connections to 43 per cent.[17]

---

16 According to one report, Digicel's share of the urban mobile market, where bmobile competes with attractive rates for data, is 60 per cent (Oxford Business Group 2020).

17 bmobile's coverage map on its website includes the hopeful proviso that the map 'may vary from time to time as coverage expands around PNG' (www.bmobile.com.pg/NetworkCoverage, accessed 17 July 2023).

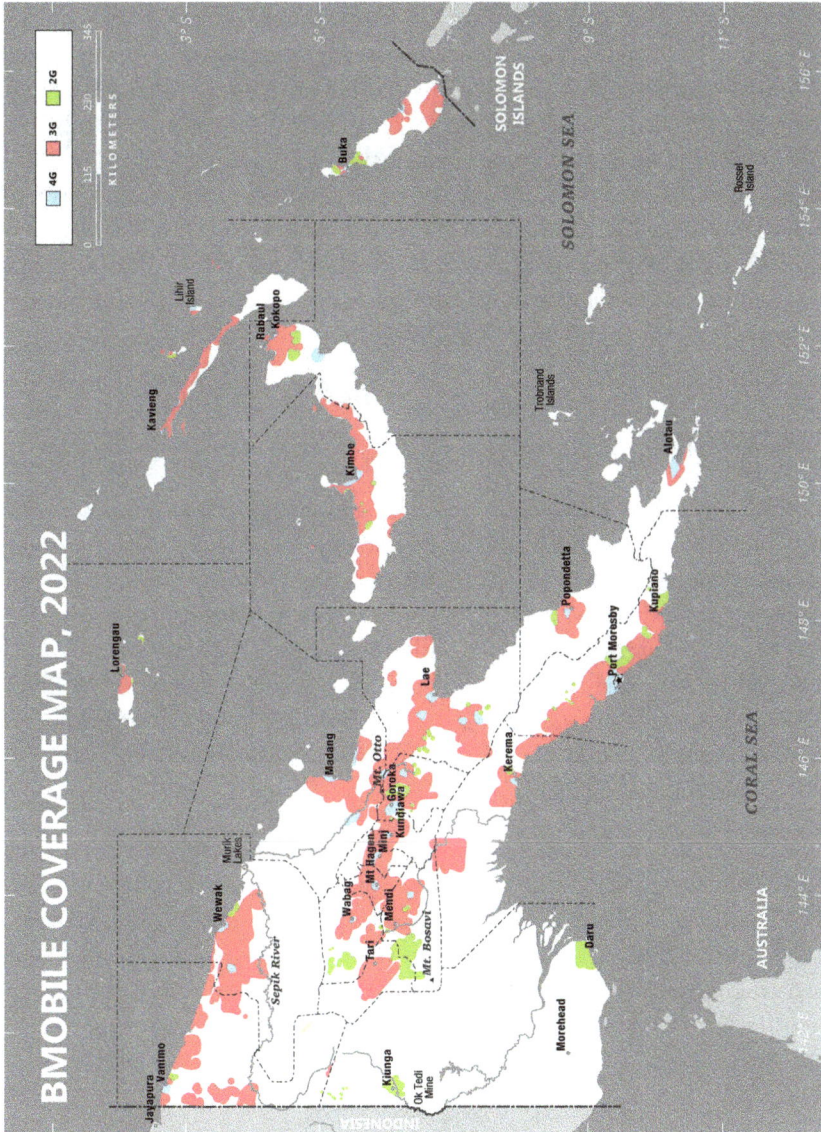

Map 1.2. bmobile 2022 coverage map with 2G, 3G and 4G zones.

Source: The Map Shop.

In short, what has emerged since the introduction of mobile broadband service to PNG in 2011 is an uneven mediascape familiar from other countries in which rural areas, provincial towns and major cities are differentiated by their access to 2G, 3G and 4G LTE, respectively (see Curry et al. 2016). The initial telecommunications democracy that Digicel established in 2007 in which all citizens were equally served, if they were served at all, by the same 2G towers has yielded to a more hierarchical arrangement in which one's location in town or in the bush effectively determines one's status in the network and perforce the quality of one's 'infrastructural citizenship' (Lemanski 2018; Foster 2023; see Conclusion).

Gateways

Despite its initial success in building a domestic telecommunications network, Digicel struggled against the monopoly on access to the gateway to international submarine cables retained by Telikom, the state-owned incumbent. Eventually, Digicel secured a gateway licence that allowed access to bandwidth via satellite—an arrangement that was less expensive than purchasing directly from Telikom. The cost of bandwidth in PNG nevertheless remained notoriously expensive until 2019, contributing to some of the highest internet prices in the world (Deloitte Touche Tohmatsu 2016; Watson et al. 2022).[18]

Access to cables became increasingly important with the spread of smartphones and the transition from voice and text messaging to data. PNG national policymakers addressed this issue by creating a new state-owned enterprise, DataCo, charged with managing the gateways and constructing a new fibre optic National Transmission Network (NTN). One significant development in this ongoing story involved the intervention of the Australian Government in the construction of a new submarine cable

---

18   Prices declined significantly in 2019 (see Watson 2020a), when rates offered by Telikom and bmobile dropped below those offered by Digicel: 'Telikom PNG reduced prices for data bundles by as much as 80% in July 2019, and Bmobile reduced retail prices for both data and fixed broadband by the same margin. Lower prices were followed by a threefold increase in traffic, leading to capacity expansion. These improvements were important in light of the global Covid-19 pandemic, which further accelerated demand as individuals adopted social-distancing measures, and shifted to working and studying from home' (Oxford Business Group 2020: 118). This reduction in costs and increase in capacity were realised mainly by customers in the urban markets served by bmobile/Telikom rather than by rural customers served by Digicel. Hence the letter from Issen N Umnbah to *The National* ('Bmobile Urged to Reach Out', 27 July 2020b) beseeching bmobile 'to extend its coverage to reach the unreached areas of Papua New Guinea'.

linking Port Moresby with Sydney. This intervention was designed to block the Chinese company Huawei Marine, which was at the time already involved in the construction of the NTN, from laying the new cable. In this way, the mobile network infrastructure in PNG has been shaped by large-scale geopolitical rivalries as well as more local considerations.

The Undersea Network

Like other Pacific Islands countries, with the important exception of Fiji, PNG is poorly positioned with regard to the undersea network of cables that enables international telecommunications, including all transoceanic internet traffic (Starosielski 2015). When Digicel began its PNG operations in 2007, the country was served by a single international submarine gateway, APNG-2, under the control of Telikom. APNG-2, a reused part of an optical fibre cable that formerly linked Australia to Guam, was put in service in 2006; it replaced APNG-1, a coaxial copper cable that from 1976 linked Port Moresby to Cairns.

In 2009, an 80 km spur of PPC-1, an optical fibre cable that connects Australia and Guam, was landed at Madang on PNG's north coast. PPC-1 offered a 10 Gbps data capacity compared with 1,100 Mbps for APNG-2. However, the inconvenient landing at Madang required the construction of a 300 km connection to the major port city of Lae by hanging fibre optic cables beneath PNG Power's electricity lines. This connection was not completed until July 2012, as part of the emergent NTN that would eventually consolidate existing and newly constructed bits of satellite, microwave and fibre optic infrastructure. The Madang to Lae connection would, in turn, connect the PPC-1 cable via an overland optical fibre link to Port Moresby. Like APNG-2, PPC-1 fell under the monopoly control of Telikom.

The *National Information and Communications Technology Act 2009* (the NICT Act) implemented in 2010 created NICTA. It also contained a few provisions with important implications for Digicel's emerging transition from voice to data. First, it instituted an open licence regime in which licensed carriers and specific service providers were able to migrate their licences and offer a variety services, including internet services. Second, it liberalised the international gateway. In July 2011, Digicel was awarded an international gateway licence in order to roll out its 3G mobile broadband service. That same year, Digicel purchased DataNets, a local internet service provider (ISP).

Finally, the NICT Act put in place a version of the NETCO/SERVCO model that had been proposed for mobile network operators but beaten back in 2007. In September 2011, NICTA recommended that 'access to international communications cables, gateway facilities and satellite links should be made "declared services," meaning PNG Telikom has to supply access to its competitors' and be subject to price regulation (Oxford Business Group 2012b). Telikom would separate its retail and wholesale activities, the latter being transferred to a new company, PNG DataCo, that would hold and operate a number of Telikom's infrastructural assets, including the NTN.

Telikom pushed back against NICTA's plans to furnish the country's ISPs with non-discriminatory access to the submarine fibre optic connections via a single wholesaler, PNG DataCo, formally established in 2014. Among other objections, Telikom claimed that 'The wholesale capacity market on submarine cable for international connectivity is a newly emerging market and should not be prematurely subject to regulation' (Telikom PNG Ltd 2012). In early 2015, Telikom CEO Michael Donnelly explained that the separation of Bmobile from Telikom in 2008 (when half of Telikom's shares in its subsidiary were sold to other investors) not only opened the door to Digicel to dominate the mobile market, but also prevented Telikom from offering services to customers who sought one-stop shopping for fixed, mobile and broadband services (personal communication, 17 March 2015). Donnelly regarded the surrender to DataCo of Telikom's cash-generating gateway assets, which he valued at PGK140 million, as 'not right' and a move that Telikom's board was obliged to resist. From this perspective, only a reunited Telikom and bmobile—a reunion that did eventually happen—could provide serious competition to Digicel.

As in 2007, different PNG state agencies were at odds with each other over ICT policy, destabilising governance of the infrastructural assemblage. In early 2017, the government announced the amalgamation of Telikom PNG, bmobile and DataCo under one name, Kumul Telikom Holdings Limited (KTHL), itself a subsidiary of Kumul Consolidated Holdings (KCH). According to its website, KCH is 'the entity which holds in trust, the Government's non-petroleum and non-mining assets' (KCH n.d.) (formerly known as IPBC and established in July 2002 under the *Independent Public Business Corporation of Papua New Guinea Act 2002*). This corporate restructuring was completed in 2021, when Telikom and bmobile were merged into a single entity and KTHL abolished (PNG Bulletin Online 2021).

Digicel's Workarounds

Digicel's strategy for getting around Telikom's control over access to the submarine network relied on satellites for voice and data traffic (as well as to provide redundancy if necessary). Indeed, the Telikom CEO who objected to surrendering his company's international gateway assets pointed out that the monopoly on the two submarine cables did not prevent competitors from buying bandwidth via satellite, the price of which had fallen due to new technologies. In July 2014, Digicel began using the trunk services of O3b Networks, whose O3b satellites hover 'closer to the earth than conventional geostationary (GEO) satellites', thereby reducing latency and increasing internet speed for the user (Barton 2015).[19] The Gerehu satellite earth station near Port Moresby, part of the NTN managed by DataCo, was upgraded to an O3b system in 2018 with the support of the Australian Government (Kumul Consolidated Holdings 2015).

Ten years after Digicel's launch, Digicel senior manager Gary Seddon reported that the majority of the company's internet capacity ran on O3b satellites; Seddon noted, however, that future demands for internet bandwidth would likely compel Digicel to incorporate 'off-island fibre optic delivery methods' (Watson and Seddon 2017). Several satellite operators were supplying PNG's internet needs in 2017. Given the large population of PNG spread across many islands and mountainous regions where extending fibre optic cables is difficult and expensive, the use of satellites is likely to persist. For example, in 2019 the ADB offered USD50 million in support of:

> Kacific Broadband Satellites International Limited (Kacific) to provide affordable satellite-based, high-speed broadband internet connections to countries in Asia and the Pacific, especially in remote areas of small island nations in the Pacific. (ADB 2019)

Kacific CEO Christian Patouraux said that 10 per cent of the company's bandwidth had been allocated for PNG, with the expectation that schools, hospitals and villages in remote and rural areas could be connected at moderate costs of about USD1,000 per month (Munoz 2020).

---

19 'O3b Networks became a wholly owned subsidiary of SES S.A. in 2016 and the operator name was subsequently dropped in favour of SES Networks, a division of SES. The satellites themselves, now part of the SES fleet, continue to use the O3b name.' From: en.wikipedia.org/wiki/O3b_Networks, accessed 28 November 2022.

Geopolitical Friction and the National Transmission Network

The creation of PNG DataCo was intended to facilitate the upgraded infrastructure required by PNG's participation in the digital world. In addition to controlling and managing the international submarine gateways, DataCo would operate the NTN, including the integrated fibre optical network stretching from Madang to Lae through the highlands and down to the coast and on to Port Moresby. Telikom had, in 2014, already begun building a microwave transport system with a trunk line to connect Port Moresby and Lae and thus significantly improve access to the PPC-1 submarine cable. The new fibre network would ensure greater stability. It piggybacked in part on pipeline constructed for ExxonMobil's liquefied natural gas (LNG) project, stretching from the Hides gas fields in the southern highlands to the LNG processing plant 50 km outside Port Moresby. This link in the NTN was completed in 2013, with the extension to Port Moresby finished in 2015.

Another major component of the NTN is the Kumul Submarine Cable Network (KSCN), completed in 2020, which provides domestic connectivity across 14 coastal provincial capitals and Port Moresby, and international connectivity by a link to Jayapura in West Papua (Indonesia). The project was financed with a USD270 million loan from China's Exim Bank (National 2020c). It was built by Huawei Marine Networks, majority ownership of which was sold in 2019 to another Chinese company.

The KSCN, like the NTN, was built in order to reduce friction in the telecommunications infrastructure—that is, to stabilise network connectivity across vast distances and under extreme environmental conditions. For example, access to the PPC-1 pipe would meet the challenge presented by the limited capacity of the aging APNG-2 pipe. But friction has not of course been eliminated entirely. Heavy rains have taken down the powerline masts supporting the terrestrial fibre link between Madang and Lae, and in 2017 an earthquake snapped the spur connecting PPC-1 to the Madang landing. In 2019, an earthquake severed links in the new Kumul cable between Madang and Lae and between Lae and Popondetta. Repairing severed undersea cables is costly and highly specialised work that only a handful of companies can supply (Starosielski 2015). Suwamaru (2020) has pointed out, moreover, that the landing points of the Kumul cable, Madang and Port Moresby excepted, are distant from the nearest mobile switching centres of service providers such as Digicel. These providers would have to build their own optical fibre interconnection to access wavelength from the KSCN, which would then be distributed to rural end users mainly through slower and less reliable radio frequency–based links.

There is more to friction, however, than extreme weather events and troubling logistics. Geopolitical as well as geophysical forces can destabilise an infrastructural assemblage. In the case of PNG, the government's financial relationship with China and its contracts with Huawei have become increasingly central to the future of telecommunications in the region. On the one hand, China has been accused of practicing a form of debt-trap diplomacy in which its largesse to developing Pacific Islands nations like PNG becomes an instrument of resource extraction. On the other hand, Huawei's extensive involvement in building critical digital infrastructure has aroused national security concerns and has even led to charges that the company serves as an agent of espionage for the Chinese Government. These geopolitical considerations have prompted Australia and the United States to assert their power and influence in shaping the telecommunications infrastructural assemblage in PNG.

***

Upon his return from Beijing in early 2009, PNG Prime Minister Sir Michael Somare expressed his appreciation of the decision of Huawei Technologies to partner with Telikom PNG on the development of an Integrated Government Information System (IGIS) (People's Daily Online 2009). Somare mentioned that the partners had previously signed an agreement worth USD3 million to improve microwave transmission. But the agreement to develop IGIS, finally launched in 2014, involved a much larger investment in the form of a USD53 million loan through China's Exim Bank.

The IGIS loan was not the last from China to PNG focused on telecommunications and digital projects. In 2013, the PNG Government borrowed USD63 million to initiate a national identity card registry and recruited Huawei Technologies to build a National Broadband Network (NBN). The NBN was to entail

> the construction of a network capable of delivering broadband via ADSL2+ technology to more than 80,000 premises across the country. Further, an additional 8,000 locations are set to be connected via Gigabit Passive Optical Network (GPON) technology at speeds of

> up to 100Mbps. According to Telikom, the NBN project will also see the creation of a new high-speed broadband backbone, with the long-haul microwave transmission network set to offer open access to both fixed line and mobile broadband operators. (Comms Update 2013)

In August 2020, State Enterprises Minister Sasindran Muthuvel noted that the PNG Government owed China's Exim Bank USD200 million for the NBN as well as the USD270 million for the KSCN (National 2020d). Given that massive investment, he asked, how could anyone object to paying Huawei an additional PGK17.67 million to integrate these assets?

As early as 2010, concerns were raised in both PNG and Australia about Huawei's involvement in the IGIS project. These concerns were voiced again about Huawei's contract for the NBN, especially given that only a year earlier Australia had blocked Huawei from bidding on contracts in Australia's own USD38 billion National Broadband Network due to cybersecurity issues. In 2018, Australia, Japan and the US exerted pressure on the PNG Government to drop Huawei from the KSCN project with an unsuccessful counteroffer. Minister for Public Enterprise and State Investment William Duma refused to undo the deal with Huawei, observing that the counteroffer was 'a bit patronizing' and that Huawei had already completed about 60 per cent of the project (Westbrook 2018).

By 2020, concerns about cybersecurity and debt-trap diplomacy had converged, heightened by a report that the data centre built by Huawei as part of the IGIS and opened in 2018 presented clear and dangerous security risks. These risks were mitigated to some extent by the fact that the data centre had fallen quickly into disrepair due to a lack of funds for required maintenance including the renewal of software licences and the replacement of batteries (Grigg 2020). But 2020 also saw the completion of one major project that signalled both the success of Australia in limiting China's growing presence in the Pacific and the way in which the telecommunications infrastructure in PNG was being shaped by geopolitical rivalries: the Coral Sea Cable System.

The 4,700 km, 40 Tbps capacity Coral Sea fibre optic cable connects Port Moresby and Honiara (Solomon Islands) with Sydney. It replaced the recycled APNG-2 cable, which was decommissioned in February 2021. Alcatel Submarine Networks laid the cable, for which a AUD136.6 million contract was awarded to Vocus Communications of Australia. The project also involved the construction of a domestic submarine network for Solomon Islands.

About two-thirds of the funding for the Coral Sea cable came from the Australian Government, with PNG and Solomon Islands contributing the remainder. The Australian Government's generosity was prompted by the announcement in July 2017 that Huawei Marine had signed a no-bid contract with the Solomon Islands Submarine Cable Company to build a fibre optic connection between Sydney and Honiara, with a further domestic extension. This contract aroused the same security concerns that led to the ban of Huawei from bidding for the Australian NBN as well as a desire to assert Australia's presence in a region more and more open to Chinese influence.

While the Coral Sea cable was successful in blocking Huawei and the Chinese state's involvement in building an international cable connection with a landing point in Australia, it was not as immediately successful in reducing the cost of internet access in PNG. Several months into operation, retail prices for internet service had not declined significantly, if at all (Watson, Airi et al. 2020). A 16 May 2020 report in the *Australian* newspaper titled 'China Fears Over Australian-Funded PNG Cable too Dear to Use' claimed:

> PNG's National ICT Authority published new pricing in December for the Coral Sea Cable of $US55/megabits per second/ month, but it's understood Data Co is refusing to pass on the new rates. It is charging wholesale rates of $US350/mbps/month for a 20mbps connection on the Coral Sea Cable, while satellite companies offer broadly equivalent packages for $US250.

Robert Potter (2021), a cybersecurity adviser, speculated that DataCo was using the savings enabled by the Australian-financed Coral Sea cable to subsidise the company's repayment of its loan from China's Exim Bank in connection with the Kumul domestic cable.

NICTA reviewed and reset pricing at the end of 2020 to PGK209 (USD59)/ megabits per second/month with further reductions planned into 2023. The question remains open, however, as to what effect new wholesale pricing, even if accepted by DataCo, will have on retail prices (see Watson et al. 2021, 2022; Watson 2022), as the wholesale cost of bandwidth is a relatively small fraction of the total cost of delivering internet services in PNG, especially in rural areas. Satellite, moreover, offers providers some protection against the infrastructural friction generated by non-geopolitical events ranging from cable-snapping earthquakes to cable-cutting vandalism.

2

Making a (National) Market: Advertisements, Promotions and Sponsorships

Introduction: Market Work

What sets markets in motion? What impels their extension and ensures multiple and repeated transactions? These are the kind of questions posed by an economic sociology that eschews abstractions and generalisations and embraces the concrete activities of particular buyers and sellers (Callon 2021). They are questions that one can ask even of a market manifestly dominated by a single actor, such as the market for mobile communications in Papua New Guinea (PNG) dominated by Digicel within a year after its launch in 2007. A GSMA report indicates that Digicel commanded 92 per cent of the mobile market share in 2018 (Highet et al. 2019). But, as more than one Digicel official mentioned to me, the company's competition was not the goods offered by other mobile network operators. Instead, the competition consisted of other consumer goods affordable to Papua New Guineans with very limited budgets—cans of Coca-Cola (Foster 2008), packets of instant Maggi Noodles (Errington et al. 2012) and, of course, unbranded but no less compelling things such as betel nut (*buai* in Tok Pisin) and tobacco (*brus* in Tok Pisin).

Markets are made and can be unmade; they do not simply spring organically out of pre-existing latent demand or respond automatically to variations in supply. Demand, after all, can be dampened as well as stimulated. A wide array of people, occupations and devices effectively work on the market, they 'construct it, move it, organize it, manage it and control it' (Cochoy and Dubuisson-Quellier 2013). In other words, as Callon (2021) persuasively argues, we need more than bloodless theories of aggregated supply and demand to see how everyday market activity is produced:

> Market activities always face the same questions: Who is ready to pay, how much, and for what? But what does change between situations are the mechanisms of solving these problems (Callon 2021: 30).

We need, then, to pay attention to how market professionals deploy a range of mechanisms or devices—eye-catching window displays or glossy brochures—in order to entangle persons and things and keep them entangled.

This chapter proceeds from the basic premises of Callon's (2021) approach to economic transactions and his unapologetically retail view of markets. It briefly considers a few, and only a few, of the innovative and aggressive ways in which Digicel attempted, with flair and success, to make a market for its goods and services in PNG. Many of Digicel's marketing strategies were pioneered in the Caribbean and exported to the Pacific. These strategies often elicited nationalist responses; that is, attempts on the part the incumbent mobile network operator to portray itself as the true representative of the national popular interest. Some of Digicel Pacific's marketing strategies suited local conventions of taste and propriety better than others.[1] For example, while the sensuality of Digicel's initial Caribbean self-presentation risked offending the sensibilities of conservative religious Fijians, the Rastafarian imagery and reggae sound of Digicel's ads resonated with urban youth in PNG (see Horst 2014). The adjustments made by Digicel's marketing managers illustrate the sort of ongoing work by which the characteristics of goods and perforce the characteristics of buyers and sellers are 'qualified' (Callon 2021: 53).

Market competition in PNG, as elsewhere, involves rival companies qualifying and requalifying themselves (and their brands) in terms that are both intelligible and attractive to local consumers (see Callon et al. 2002). I have discussed elsewhere (Foster 2008) the dynamics of qualification with respect to the marketing of soft drinks, noting that a product like a can of Coca-Cola is endowed with meaning—qualified—at every moment

---

1   In 2008, chiefs in Bougainville demanded that Digicel remove a billboard erected at the Buka airport that depicted women stepping on an *upe*, a woven pandanus hat worn by young boys undergoing traditional initiation rites (Laukai 2008). The *upe* is featured on the Bougainville flag and PNG stamps.

of its life: design, manufacture, marketing, use, disposal, recycling and so forth. These qualifications by different agents who most likely will never encounter each other face-to-face or even know of each other's existence are not and need not always be identical or even be compatible with each other. Callon et al. (2002: 201) rightly observe: 'There is no reason to believe that agents on the supply side are capable of imposing on consumers both their perception of qualities and the way they grade those qualities.' Indeed, entangling persons and things is never guaranteed. There is an inherent tension between a supply-side agent's desire to render the consumer's purchases routine and the risk posed to the supply-side agent in so doing of losing touch with the motivations of a consumer whose behaviour has become unreflective.

I consider in this chapter how the marketing strategies of mobile network operators have aligned with or against the expectations and sensibilities of the so-called gift economy, often regarded as characteristic of Melanesian societies, a theme resumed in the following chapter. That is, to what extent did Digicel, specifically, invite mobile users to imagine themselves engaged in a gift relationship with the company? I discuss advertisements first but focus mainly on two market-making activities: promotions and sponsorships, examples of market devices that 'aim to capture a potential customer's attention' (Callon 2021: 59). This focus accordingly foregrounds one particular aspect of the moral economy of mobile phones: the relations between companies and consumers. How did Digicel's marketing team fashion this relationship as reciprocal and ongoing? How did Digicel's users respond? I devote most attention to the years right after Digicel's arrival in PNG, when the company was resolutely set on expanding the customer base for its 2G network of voice and text messaging services.

Advertisements

Wherever it landed across the Pacific Islands, Digicel arrived with a big marketing splash. Towns were covered with billboards and banners, newspapers carried advertisements daily, and shops were painted in the company's colours of red and white (see Willans et al. 2022; 'Ofa 2011: 124). Flashy youth-oriented strategies developed in the Caribbean were exported to the Pacific Islands, where they met both positive and negative reactions. Digicel responded to these reactions, modifying its appeals to fit local conventions and counter the responses of its competitors. Digicel's recurring challenge was to display its commitment to the nation despite its status as a privately owned foreign company.

Digicel organised an exuberant, upbeat video to mark the company's launch in PNG.[2] The video features shots of hard-working employees and wide-smiling consumers, and even a hand-waving Denis O'Brien arriving at the Port Moresby airport—all set to the stirring music of Irish rock band The Cranberries. Scattered across the cheerful montage are clips of the various marketing activities undertaken by Digicel immediately before and after the July 2007 launch. Billboards are erected; banners are hoisted on power poles; public buses are painted red and white; retail stores, also painted red and white, are opened to waiting crowds; concerts and fashion shows are hosted. Similar strategies transformed the urban visual landscape when Digicel arrived in Samoa, Tonga and Vanuatu (Willans et al. 2022). The PNG launch video concludes with its only understatement, the proclamation 'Digicel Has Arrived'.

Digicel's aggressive marketing tactics caused trouble in both Port Moresby and Lae, where the company was instructed to pull down pennants that were put up without the official permission of local authorities. In addition to claiming urban public space as Digicel space, the company asserted a steady presence in PNG's two main newspapers, the *Papua New Guinea Post-Courier* and *The National*. These papers circulate mainly in the cities and towns of PNG, where copies are often passed from one reader to another, but they also make their way into rural areas where they find a readership hungry for news about current affairs and sports.

Every day in July and August 2007, red and white advertisements, often full-page, proclaimed the company's arrival and intention to stay despite government interference (see Chapter 1) along with its promises for cheaper rates, inexpensive handsets and more extensive coverage. Digicel tracked its progress in opening stores and extending services outside Port Moresby and Lae—to Mt Hagen in the highlands and Kokopo in the islands, and to locales that never before had mobile service. The ads conjured an expanding and inclusive national space-time, an imagined community in which all members could connect with each other across long distances (see Foster 2023).

---

2   I am uncertain about where the video was shown in PNG at the time of the 2007 launch, but it was posted to Digicel PNG's Facebook page in 2020 for the company's 13th anniversary and as of this writing is available on YouTube: www.youtube.com/watch?v=dqNFTrMv5Q0, accessed 4 July 2023.

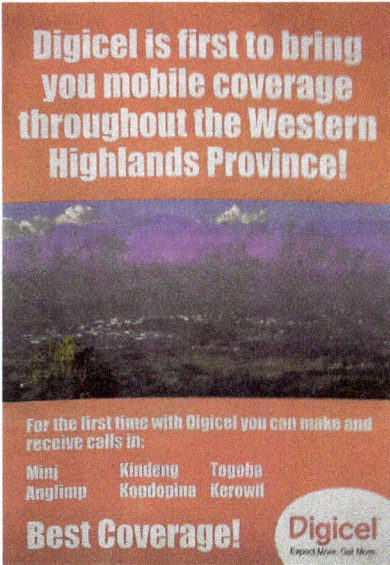

Figure 2.1. Digicel newspaper ad (Western Highlands coverage).

Source: *The National*, 20 September 2007.

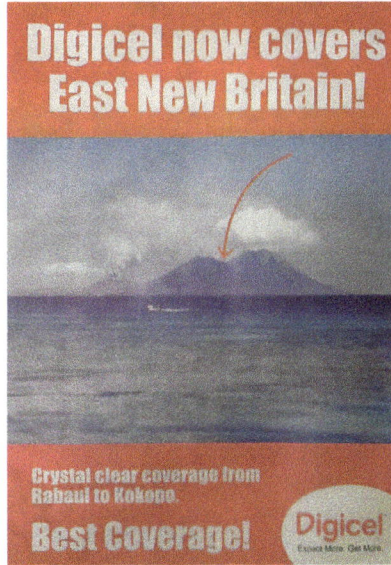

Figure 2.2. Digicel newspaper ad (East New Britain coverage).

Source: *The National*, 3 October 2007.

This community, moreover, was highly egalitarian. Digicel had, unlike its rival BeMobile, put handsets and airtime within reach of rural as well as urban consumers, little guys (the so-called grassroots) as well as big shots. As one letter writer to the *Post-Courier* succinctly put it: 'Digicel is a grassroot company which knows the needs of the simple grassroot people of this nation. B Mobile is only for the rich' (30 July 2007l).

The marketing response of Telikom to Digicel's populism (and popularity) was perhaps as predictable as it was inevitable. A newspaper statement protesting Digicel's launch revealed how Telikom would play the nationalist card: '**Telikom is owned by the people of Papua New Guinea**. Telikom will not be bullied by a foreign company that ignores government policy …' (National 2007d, original emphasis). The statement baldly appealed to economic nationalism and asserted that all of Telikom's profits 'stay in PNG'.

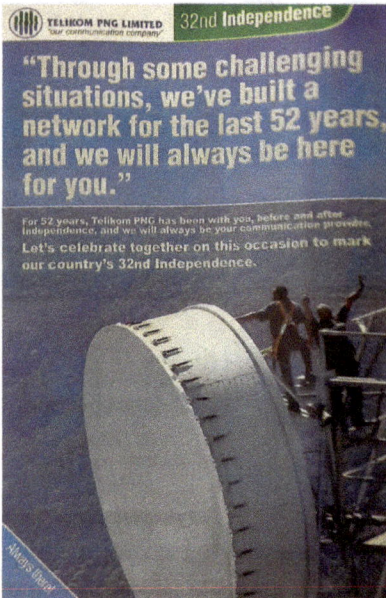

Figure 2.3. Telikom newspaper ad (Independence Day).

Source: *The National,* 13 September 2007.

Telikom subsequently represented itself in ads that were part of a concerted 'rebranding' as the national carrier, using the tagline 'Always Here!' to remind folks that the company had long been there to serve the people. Hence the message of one full-page ad, appropriately placed on Independence Day weekend in September 2007, that pictured two Telikom workers surveying the mountain valley below from atop a dizzyingly high tower: 'Through some challenging situations, we've built a network for the last 52 years, and we will always be here for you'.

The rebranding exercise acknowledged past shortcomings such as slowness in repairing faulty landlines, regarding which one company representative commented: 'We say thank you for your patience and continued loyalty to Telikom.' As then CEO Peter Loko put it, he wished consumers to see Telikom 'for what we are today, not what we might have looked like earlier' (National 2007e).

Telikom publicised a steep reduction in the price of its BMobile 'start-up kit' of a SIM card and credit from PGK125 to PGK25—a move that was met with scepticism as well as enthusiasm. Letter writers to the national newspapers noted that Telikom only reduced its price in response to competition from Digicel and, therefore, was previously guilty of gouging consumers. At least one writer pointed out that the previous price of 125 kina included 100 kina in 'free' credit, whereas the new price of 25 kina included 10 kina of credit, making the reduction in price of the SIM card and connection only 10 kina. Telikom also promoted its coverage in rural areas, where Digicel was rapidly deploying its network of towers. The text of one ad declares: 'We Are Reaching Out to the Remote Parts. And It's Your Brother, Mother and Relatives That We Thought About in the Village. Always PNG, Always TELIKOM, Always There.'

Figure 2.4. Digicel newspaper ad (Swap SIM cards).

Source: *The National*, 21 September 2007.

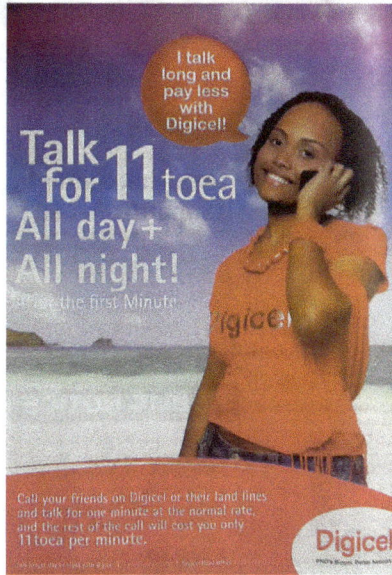

Figure 2.5. Digicel newspaper ad (11 toea talk).

Source: *The National*, 3 December 2009.

Digicel's advertisements fashioned the company not only as a saviour in delivering services to long neglected people and places, but also as an accessory to the pursuit of a modern and urbane lifestyle that appreciates the value of traditions. Young men with long dreadlocks and young women in T-shirts and jeans brandished phones and smiles in advertisements for discounted rates or special deals.

At the same time, images of youths and adults dressed in recognisably traditional gear adorned not only print advertisements but also the scratch-off 'flex cards' used to sell prepaid airtime (see Figure 3.1, Chapter 3). Digicel thus replicated a marketing strategy previously used by other vendors of everyday consumer goods such as soft drinks and tinned meat, in which Papua New Guinea was imagined as both traditional and modern, rural and urban; and in which the visual juxtaposition of a new imported technology and symbols of timeless homegrown custom illustrates the best of two worlds (Foster 2002).

***

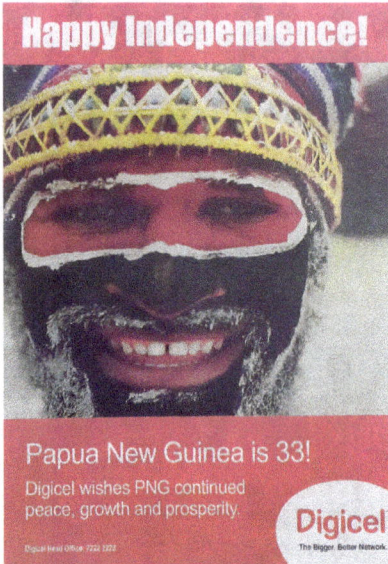

Figure 2.6. Digicel newspaper ad (Traditional dress Independence Day).

Source: *Post-Courier*, 16 September 2007.

Figure 2.7. Digicel newspaper ad ('Call Me' service).

Source: *The National*, 27 August 2007.

The making of a mobile phone market in PNG entailed more than the proliferation and accumulation of momentary transactions between buyers and sellers. From the very beginning, Digicel's marketing directly engaged the people who used its products and services in an open-ended durable relationship. The company introduced the free 'Call Me' feature, as it did in the Caribbean (Horst and Miller 2005: 762), as a gesture of recognition that people sometimes run out of credit and need to ask someone else to make the call.

Similarly, the company announced that it would not charge users when they retrieve voicemail. One could argue, of course, that these initiatives were mainly intended to promote usage. But the point is that the company engaged users in what appeared to be more like a long-term gift relationship than a one-off commodity transaction—a gesture even more explicit in the company promotions discussed in the next section of this chapter.

Figure 2.8. bemobile newspaper ad (bemobile Orange Men).

Source: *The National*, 11 December 2009.

Digicel's marketing presence in newspapers and on public signage was impressive in the first few years after the company entered PNG, perhaps peaking in 2009 when a paint war broke out between Digicel and the rebranded bemobile, which had claimed the colours orange and purple to oppose Digicel's red and white (see Watson 2011: 226). bemobile used the 'bemobile Orange Men' in print ads, which harkened back to the use of similar orange men (portrayed by the well-known Blue Man Group) to promote Mirinda brand orange-flavoured soft drinks in the 1990s.

The rival companies painted bus sheds and storefronts to such an extent that complaints began to appear in newspapers about the unsightly look of Port Moresby's public space.

Telikom briefly entered the quest to qualify itself with a distinctive colour combination in 2011, when its short-lived CDMA mobile service began in Port Moresby and Lae. The slogan 'Say Yello with Citifon' signalled the brand's appropriation of the colour yellow (along with a complementary purple). Citifon took out bright yellow full-page newspaper ads for promotions that included a limited-time tripling of credits when topping up and a contest in which a daily call was made to a lucky Citifon number during the FM100 Drive Time Program: 'Answer your phone Saying YELLO and WIN Cash' (*Post-Courier*, 4 May 2012a). Telikom, whose green and blue colours figured less prominently in its marketing, advertised a monthly contest for a 'dream car' worth PGK25,000. Entering the contest required one to purchase prepaid top-ups of PGK50 (for Citifon, fixed wireless or data accounts) within a month; additional PGK50 top-ups resulted in additional entries (see next section for more on promotions).

By 2018, Digicel's presence in both daily newspapers and in Port Moresby's public space was much more subdued. On the one hand, the company had established its dominance over the market it had effectively brought into being. On the other hand, the rapid shift from basic phones to smartphones enabled the company to conduct its advertising campaigns and branding exercises online, on its Facebook page and through its My Digicel app (see Chapter 4). For example, Digicel sought to exploit the affordances of smartphones to send personalised 'hyper relevant, real time messages that increase revenue, engagement and retention among app users' (SWRVE 2020). What was once a public spectacle in cities like Port Moresby and Lae had evolved into the private communions that occur during the many times every day when people swipe and gaze at their touchscreen.

Promotions

Promotions, actions intended to increase sales and to raise awareness of a good, illustrate well Callon's (2021: 59) idea of a market device that aims 'to capture potential consumers' attention and to make them deviate from their original trajectories'. Digicel extensively used promotions—discounts, giveaways, contests and so forth—in all of its Pacific markets; B Mobile also used promotions in PNG (Vodafone Fiji rolled out promotions frequently and since 2022, Vodafone PNG has continued this practice in PNG). Promotions shaped the behaviour of individual consumers, for example, influencing when and how consumers recharged (or topped up) their accounts. On the one hand, these efforts seemed to construct a gift relationship between consumers and companies in which loyalty was rewarded with occasional perks and bonuses. On the other hand, these efforts sometimes pushed the boundaries of conventional morality, as when advertisements for phone lotteries in PNG exposed Digicel to accusations of promoting gambling. Promotions thus provide a handy means for apprehending the dynamics of consumer–company relations within the moral economy of mobile phones.

***

Foreign visitors to marketplaces in Papua New Guinea are often struck by one fact in particular: there is hardly any price competition. One can wander through the rows of betel nut vendors and find the same number of similar-sized nuts for sale at the same price irrespective of who is selling them. The same goes for sweet potatoes or tomatoes. Noisy higgling and haggling is distinctly inappropriate in these marketplaces. Buyers make their choices on the basis of the quality of the merchandise and/or the quality of the social relationship they enjoy with particular vendors. Indeed, it is not uncommon for vendors to add a gift of something 'extra' to a transaction as a sign of the ongoing relationship that the transaction performs and reproduces (Busse and Sharp 2019). The PNG marketplace is no place for the cold, impersonal market.

From the perspective of PNG marketplaces, the boisterous price competition that Digicel sprang on BMobile in 2007 was as highly unusual as it was eagerly welcomed. Digicel heavily subsidised handsets, lowering the price of a basic phone at one point to PGK19 (USD6.33). The pressure of competition by 2009 led bemobile to offer a handset for PGK29 that came with 5 minutes of free talk time, five free SMS and double credits on the first top-up (the price dropped to 19 kina for the day of Christmas Eve only).

Digicel not only lowered prices, but also offered two-for-one deals that caused long lines to form at retail outlets.

Figure 2.9. bemobile newspaper ad (19 kina Christmas Eve handset).

Source: *The National*, 24 December 2009.

Figure 2.10. Digicel newspaper ad (2 for 1 handset offer).

Source: *The National*, 27 July 2007.

The strategy here was hardly novel: subsidise the razor so that profit can be made from recurring purchases of blades. But the purchase of one razor does not stimulate another person's use of blades, whereas the purchase of two handsets for the price of one doubles the number of users with an incentive to purchase prepaid credit for making calls or sending text messages. Digicel thus steadily grew what it would eventually proclaim its 'bigger and better network'.

The sheer audacity of Digicel's competition with Telikom/Bemobile was epitomised by the bold SIM-swap offer which Papua New Guineans would often spontaneously recall in conversations with me 10 years later (see Figure 2.4). Newspaper ads for this offer exhorted people to 'Join Digicel for Free!' The ads instructed: 'Bring in your B-Mobile SIM card and swap it for a FREE Digicel SIM card and get K20 for FREE!' This promotion was successful enough to compel Telikom to match it, urging consumers to swap Digicel SIM cards for BMobile SIM cards. (This move, however, like the reduction of prices in start-up kits, again left Telikom vulnerable to accusations of being an unoriginal copycat.) Indeed, the intensity of the competition between Digicel and BMobile was widely suspected to account for the uptick in vandalism targeting Telikom cables in Port Moresby around the time of Digicel's arrival (National 2007f). The late Peter Loko, then CEO of Telikom, admitted that he himself suspected the involvement of Digicel's agents in the sabotage (personal communication, 1 August 2015).

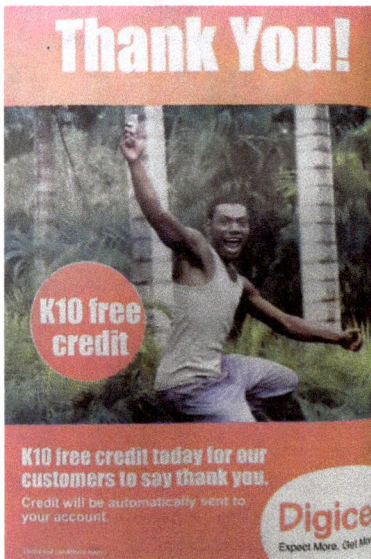

In other respects, Digicel's marketing strategies seemed to align with the moral expectations of a gift economy. For example, the company placed a full-page newspaper ad just days after launching in PNG with the message of 'Thank You' and the gift of PGK10 phone credit sent automatically to the accounts of all customers.

Figure 2.11. Digicel newspaper ad (Thank You 10 kina free credit).
Source: *The National*, 27 July 2007.

Such gifts, of course, stimulate phone use and thus serve the interests of a company motivated to socialise users new to the possibilities of telecommunications. The logic is explicit in many of Digicel's promotions, such as the August 2007 offer to 'talk free for 30 minutes' after topping up one's account for any amount. The free 30 minutes functions—much like the 'extra' betel nut or banana in a marketplace transaction—to embed the purchase of credit in a durable, open-ended relationship, in this case, between corporate and natural persons. These 'gifts' often shaped the ways in which prepaid customers would organise their phone use, a topic taken up in Chapter 3.

Box 2.1

Digicel Promotion Texts, received July 2014

WIN a 32-INCH TV EVERYDAY! Simply txt ur lowest bid to 16783 e.g. txt 16 if u want to bid 16t or 246 for K2.46. Auction closes 5PM daily. 49t/txt

Play Guardians of the Universe! Text GUARD to 204! Win up to K70,000 this July! Rank 1 wins K36,000, Rank 2 – K25,000 & Rank 3 – K10,000. U must be 18+ to play

Text ALL Day! Text SMS to 1629 & get 50 Digi to Digi SMS for 75t, valid till midnight. Dial *130# for bal

DOUBLE INTERNATIONAL MINUTES! Enjoy 120 mins of calls to Australia, NZ, China, India, Indonesia, Malaysia at K12! Text CALL to 16546! Offer expires today!

Switch to International Pass! Call AUS/NZ/China/India/Indonesia/Malaysia/Philippines for 30 mins at K10! Txt to 16545!

July Travel Promo! Travel in July & win a holiday in Cebu or Bali. Text your name, location, ticket number and email address to 71004414. Think PX! Fly PX!

Switch to International bundle! Call AUS/NZ/China/India/Indonesia/Malaysia for 60 mins at K12! Txt to 16546!

Get DOUBLE the WEEK DATA PASS Today! Text WEEK to 1634 now & get 300MB for K10 valid for 7days.

You received K1 Bonus Credit valid for 24hrs. Free credit valid for Calls & SMS to Local numbers only. Dial *130# check balance. Thank you!

Send 1 more sms and enjoy 1t SMS all day.

Get DOUBLE the MONTH DATA PASS Today! Text MONTH to 1634 now & get 1.8GB for K65 valid for 30days.

Thank you for joining Digicel. You have been awarded a free internet pass of 5MB, valid for a week.

During the month of July 2014, I received many text messages from Digicel (see Box 2.1). Some of these texts offered gifts: 5 megabytes (MB) of data or 1 kina of credit. Other messages offered discounted rates that were available for a limited time, while yet others solicited entries for contests. Like giveaways of T-shirts and other swag, contests have been a visible element of the marketing mix in PNG for all sorts of consumer products, from soft drinks to rice (see Foster 2008). In the weeks after Digicel's arrival it was thus not uncommon for the makers of local products (such as Ocean Blue Mackerel) or local stores (such as City Pharmacy) to sponsor contests in which mobile phones were the main prize. Digicel itself organised contests for big-ticket items such as a Toyota Land Cruiser and a house.

In 2009, in the weeks leading up to Christmas, bemobile sponsored a contest in which 100 pigs were given away. Buying a bemobile phone in December and making a call were required for entry in the draw. Pigs are valuable wealth items in PNG that circulate in marriage payments and compensation payments for homicide. They are also important items in feasts associated with funerals. The pig contest made cultural good sense.

Figure 2.12. bemobile newspaper ad (Win a pig contest).

Source: *The National*, 9 December 2009.

Mobile users often complained to me about the frequency with which they were targeted with promotional texts, and they also shared other concerns about these enticements. Questions were raised about the morality of certain promotions in PNG and Digicel's moral obligations to its customers. For instance, the text message inviting me to 'Play Guardians of the Universe' came with a notice that only adults 18 and older were eligible to play. The notice suggests that this game might be more like gambling than a prize contest. And, indeed, the game brought to the surface latent tensions within the moral economy of mobile phones—tensions between companies and state agents as well as between companies and consumers.

In 2010, public outcry led within two weeks to the suspension of a mobile phone lottery that had been introduced by a firm called PNG Lotto in partnership with Digicel. The lottery required entrants to send a text message to a special number indicating the entrant's choice of six numbers between 1 and 44. Each entry cost PGK4.00 (USD1.60) and came with a receipt advising that players must be 18 years or older. The first draw was set at PGK1 million. The Catholic Church, the PNG National Council of Women and at least two members of parliament (MPs) protested that the lottery would put children at risk. Andre Mald, then MP for Moresby Northeast, insisted that it was necessary for government to regulate mobile phone companies and the services they provide (Post-Courier 2010).

Box 2.2

Letter to the Editor, *Post-Courier*, 27 April 2012

Digicel, Please make winners known to public

I express concern about text messages I am receiving from a mobile number '200' which reads 'text FISH to 204 for a chance to win K20,000! 49t per text.'

I tried texting that and got a reply saying I was ranked 569th.

And thank goodness if I want to be ranked number one I would have to send about 200 or more text messages. 200 X 49t will equate to K98. I have tried the game once.

I was on the top five rank and tried my best to get to first but could not.

After buying and texting so many K20 worth of messages while standing at the glass counter of an Asian shop and ordering flex cards until the shop ran out of K20 flex cards and I also realized that I have spent so much in the quest to be ranked 1 which I never did achieve so I gave up.

This makes me wonder how much money the person on top rank is spending on flex cards to be on top and who that person is. Is it really a person?

And did the person win and get the money eventually?

There should be some advertisement in the media to show that somebody has won. I'm just wondering who is winning all these money because the text message is almost popping up on my mobile everyday and I just ignore it.

Please, can Digicel verify this mobile lottery competition and if people are winning.

Also if somebody has won, please write and let people know that this mobile lottery is for real.

Or is it some kind of technique to lure customers to spend more on flex cards.

Concerned Flex Hunter,
Goroka
EHP

While a mobile phone lottery has not been established in PNG, SMS games such as 'Guardians of the Universe' continued to exist after the controversy in 2010. For example, in 2017, Digicel kicked off the new year with the SMS game 'Happy Resolutions'. By sending a text for 59 toea, a player receives a token worth from 10 to 100 points. The player who collects the most points by the end of the month-long competition wins PGK70,000. The game is not a lottery, but neither is it a contest in which one enters for free and the prize winner is chosen randomly.

SMS games have periodically attracted the same intensity of concern as that attracted by the aborted mobile phone lottery. In an April 2015 letter to the *National* titled 'Digicel, Stop Sending out Push SMS for Silly Games', Romney Youriku aired several complaints. While these complaints focused mainly on the inconvenience of regularly receiving unwanted texts, questions of transparency and children's participation were also raised. In 2017, when the games again had become a matter of public concern, the commissioner and CEO of the Independent Consumer and Competition Commission (ICCC), Paulus Ain, stated that mobile network operators were not entitled to run games and 'only should be providing call services or internet services and what is prescribed under your license' (Haihuie 2017). Digicel responded by claiming that it was 'compliant with the terms of and conditions set out under the provisional licence granted by the board of the National Gaming Control Board' (ONE Papua New Guinea 2017). The National Information and Communications Technology Authority (NICTA), for its part, reported complaints from Digicel subscribers about underage gambling and bombardments of unsolicited text messages, but professed an inability to assert control given that Digicel apparently had a licence from the National Gaming Control Board (NGCB) (National 2017a).[3]

SMS games bring into the open tensions that inhere in the moral economy of mobile phones. Companies, consumers and state agents have not always been on the same page about market devices such as this particular promotion. While some consumers objected to the SMS games, others clearly were willing to pay the fee of 49 or 59 toea per text to enter. While one state agency (NGCB) deemed the games permissible, another state agency, the ICCC, did not, and a third agency (NICTA) seemed unsure about how to proceed. This divergence among different state agencies

---

3   According to one report (National 2009), the NGCB had objected in 2009 to Digicel's high-stakes competitions for homes, vans and cash prizes which the board considered an unlicensed lottery.

recalled the original split between the ICCC and the Papua New Guinea Radio Communications and Telecommunications Technical Authority (PANGTEL) that enabled Digicel to commence operations in 2007. That is, the lack of consensus among PNG state agencies enabled Digicel to avoid regulatory constraints.

In the end, Digicel prevailed and the games were not banned. However, subscribers were offered the option both to bar a mobile number from playing SMS games and to block all promotional text messages by texting DND (do not disturb) to Digicel. These provisions addressed the two main complaints of Digicel subscribers while preserving the option to play, thereby reframing the playing of SMS games as a moral choice of individual subscribers rather than a moral failing of either the government or Digicel.

Sponsorships

Digicel very quickly moved in PNG and elsewhere in the Pacific to establish its place in the life of the nation through sponsorships (for Fiji, see Horst 2018; for Samoa, Tonga and Vanuatu, see Willans et al. 2022). In fact, Digicel entered into a sponsorship agreement with Fiji's national rugby team before the company had actually launched its operations (Horst 2018). In PNG, Digicel assumed sponsorship of the championship for the domestic rugby league, even though naming rights sponsorship of the PNG team that plays in an Australian rugby league (PNG Hunters or SP Hunters) remained in the hands of the country's beer producer, South Pacific (SP) Brewery. Digicel sponsored not only sports teams, but also cultural festivals, thereby associating itself with the value and values of cultural diversity and ancestral traditions.

Is sponsorship, an activity engaged in by marketing professionals (sponsorship managers), an effective market device shaping the choices of consumers? Is sponsorship, put differently, worth the cost? Apparently, even marketing professionals don't know. It is difficult to measure the financial value of sponsorship investments in the absence of 'agreed upon measurements and financial standards for evaluating sponsorship ROI [return on investment] set by the FASB [Financial Accounting Standards Board], ISO [International Organization for Standardization] or the marketing industry' (Diorio 2020). The situation is comparable to that involved in assessing the market value or price of a brand, for which there is at least an ISO-approved market device (Foster 2013). Sponsorship seems to serve several different goals, including enhancing corporate image, generating goodwill among opinion makers and promoting awareness of brands (Dolphin 2003).

Digicel engaged in sponsorships, whatever their economic rationality, soon after launching in PNG. In September 2007, Digicel advertised its sponsorship of PNG's national soccer team, the Pukpuks, and celebrated the team's success at the South Pacific Games. The company also advertised its sponsorship of the Hiri Moale festival, a long-running cultural event held annually in Port Moresby in conjunction with Independence Day weekend.

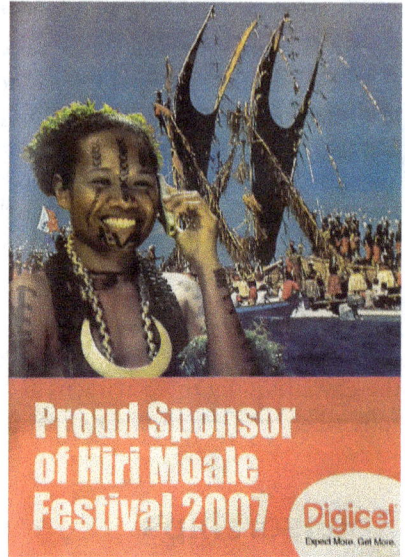

Figure 2.13. Digicel newspaper ad (Hiri Moale sponsorship).
Source: *The National*, 12 September 2007.

The symbolic intent was overt: Digicel supported the nation that its telecommunications network promised to unite. Sponsorship functioned as an effective communicative and practical means for a privately owned foreign firm to localise itself as a generous national citizen. Hence, perhaps, the greater weight given to sponsorship in Digicel's marketing mix compared to that of Telikom and bmobile.[4]

Digicel's sponsorships fell into the two main categories of sports and culture. In PNG, there is one sport that matters most: men's rugby league. Accordingly, Digicel's main investment in sponsorship was with the Papua New Guinea Rugby League. Digicel has been a 'platinum sponsor' and 'exclusive telecommunications sponsor' of the Papua New Guinea SP Hunters, a team that plays in Queensland's Intrust Super Cup competition. The naming rights sponsorship for the Hunters resided with South Pacific Brewery. However, in 2011 Digicel acquired naming rights sponsorship of the semi-professional rugby league competition played in PNG, since then known as the Digicel Cup. Digicel took over sponsorship from Bemobile (2009–11), which had taken over sponsorship

---

4   Vodafone PNG, which launched in 2022, quickly added sponsorship to its marketing mix, along with contests, promotions and giveaways of the same sort that Vodafone Fiji operates.

from SP Brewery (1990–2008). In 2015–16, this sponsorship was worth PGK2.4 million; SP's four-year (2017–20) naming rights sponsorship of the Hunters was worth PGK7.6 million. Digicel's television services are the official home for broadcasts of Digicel Cup and Hunters games as well as Australian National Rugby League games, including the annual State of Origin matches—crucial content for drawing viewers in a rugby-mad country.

Digicel's sponsorship of culture encompassed art forms that are regarded as ancestral and traditional as well as forms styled as new and contemporary. The former category consists of the various annual cultural festivals, such as Hiri Moale, that occur throughout the country each year. Among the more prominent festivals are the Goroka Show and the Mt Hagen Show, multi-day, multiethnic gatherings of performing groups wearing traditional dress. The latter category includes youth-oriented festivals such as the 2021 Milne Bay Out Loud Music Festival in Alotau, which featured local artists. Although music sponsorship was not as visible in PNG as it has been in Digicel's Caribbean markets (Horst 2014), the company certainly promoted consumption of local music through its D'Music app.

Digicel's sponsorship of Papua New Guinea's national patrimony also extended to nature. Starting in 2013, Digicel offered support in the form of in-kind services to the popular Port Moresby Nature Park, formerly known as the National Capital Botanical Gardens. According to a 2014 newspaper report, the company offered a sponsorship package that involved 'provision of free internet service, a closed user group (CUG) mobile plan for the Nature Park staff and the resale of electronic flex valued at K34,000' (National 2014a). In this way, Digicel has associated itself not only with the famously diverse cultures that comprise PNG national culture, but also with the nation's flora and fauna, for which the company assumes responsibility as an ecological steward. Sponsorship, like the work of the Digicel Foundation discussed in Chapter 6, functioned to publicise Digicel's corporate interests as PNG national interests and its investments as evidence of its good citizenship.

Part II.
Consumer Uptake:
Freedom and
Constraint

3

Mobile Economies: Prepaid, Gift and Informal

Introduction: Appropriation and Technology Evolution Cycles

The ethnographic study of mobile phones has given prominence to the question of appropriation—that is, to the ways in which people take up and use new technologies. Appropriation, however, implies more than simply adopting a new technology; it refers, in particular, to how users engage new technologies in ways neither anticipated nor imagined by providers:

> Users may modify a device, download or program new applications, invent unintended uses, or develop new practices that leverage its possibilities. This experimentation constitutes a re-negotiation of the power relationships embedded in the technology. This creative re-negotiation is the core of appropriation, the process through which users take something external and make it their own (Bar et al. 2016: 619).

Sometimes, the results of such experimentation 'may not be necessarily congruent with the provider's interest' (Bar et al. 2016: 626), in which case providers might attempt to 'repossess' the technology, for example, by coopting the innovations of users. Appropriation is thus one moment in a cycle of 'technology evolution' (Bar et al. 2016) that includes adoption and repossession.

In the case of mobile phones, the technology cycle charts the 'ebb and flow of control between users and providers' (Bar et al. 2016: 625). Put differently, the technology cycle broadly understood defines a site of negotiations between consumers and companies—not always made explicit and not always finally resolved—within the moral economy of mobile phones. These negotiations often involve companies putting constraints on the liberating appropriations of users; but in some instances, such as the backlash against mobile phone gambling (Chapter 2), they involve users resisting the enticements of companies to consume more freely.

In this chapter, I consider how mobile users experience the pecuniary dynamics of freedom and constraint on a daily basis. The spread of mobile phones throughout the Global South was made possible by prepaid subscription, the pay-as-you go system that allows people with limited financial means to purchase small amounts of credit as needed. While access to mobile telephony offered people a host of new communicative possibilities, managing prepaid subscriptions on tight budgets presented a host of new challenges. These challenges were often met creatively by clever workarounds that enabled the use of mobile phones rent-free; they were also met by self-imposed forms of discipline through which users sought to enhance their capacity to take advantage of the phone's many affordances— voice calls, text messages, web browsing and so forth. Such fiscal discipline sometimes rubbed uneasily against local norms of reciprocity that informed the circulation of phone credit as part of a distinctive Melanesian gift economy. Credit requests from kin, friends and ethnic associates (*wantoks* in Tok Pisin) thus often confronted users with a difficult moral choice.

In addition to locating the place of mobile phones in prepaid and gift economies, I discuss the informal economy brought into being by and for the expansion of Digicel's market in Papua New Guinea (PNG). This informal economy supported people's efforts to use their phones and to keep them usable under trying material conditions. Street vendors in towns throughout the country sold airtime in the form of scratch-off vouchers or 'flex cards' that were often resold several times, thus finding their way into rural areas.

Figure 3.1. Digicel 5 kina scratch-off voucher or 'flex card'.
Source: Photo by M Boie.

Self-taught repair specialists offered solutions to cracked screens and other problems, and store owners with access to the electrical grid offered a way to keep phones recharged for a fee. This informal economy in PNG, at least, is perhaps as significant as the oft-mentioned but not often achieved effects of mobile phones to spur entrepreneurship and otherwise add to the annual GDP of developing countries. The advent of smartphones in PNG, however, threatened to displace the informal economy that had grown up around mobile devices, moving 'top-ups' of account balances online and making repairs too complicated for many self-taught technicians.

Prepaid Economy

Workarounds furnish mobile users in the Pacific Islands, as elsewhere, with the means to overcome constraints on freedom to communicate. People develop workarounds to secure access to mobile communication under the best financial terms, including rent-free use of the mobile operator's network. Mary K Good (2012: 175), in her ethnography of youth culture in Tonga, provides one such example of creative appropriation. Good learned how it was possible:

to change the settings on the phone from one of the two mobile companies in Tonga so that text messages from the other mobile provider are received for free and do not take away from the card balance. If two people each had phones from both providers, then they could text back and forth for free, sending from one carrier and receiving on the other.

This happy arrangement enabled Good's interlocutor 'to do a date' cheaply and easily via text messaging with all her boyfriends. Unfortunately for many Tongan youth, the mobile providers eventually resolved the network issue that made such communication possible such that 'users could no longer "*tex ta'etotongi*," or send text messages for free' (Good 2012: 177). Rather than accommodate or coopt their consumers' practice, the companies simply suppressed it, thereby ending this particular technology cycle.

Other workarounds, by contrast, have yielded responses and counter-responses that reveal the unfolding of a longer cycle of technology evolution. Take, for example, the missed call. 'Let the phone ring twice to let us know you got home safely.' That's what my parents would tell me at the end of every visit. By ringing the phone twice and 'hanging up' or disconnecting, they would know who was calling and I would not be charged for a toll call. The arrangement allowed for sending comforting messages for free by way of deliberately missed calls.

I doubt that my parents invented this practice. It is an old example of creative appropriation. So, it could not have come as a great surprise that a similar practice—called 'flashing' or 'beeping'—was reported by early observers of mobile phone use in the Global South (Donner 2008). Flashing or beeping occurs when a user calls another phone and disconnects before the call is answered, leaving the caller's number as a trace on the recipient's mobile. The flash or beep might be an invitation for the recipient to call back (and thus bear the charge of the call) or a prearranged signal or just a friendly hello. In any case, flashing and beeping enables communication to proceed without paying rent for use of the network.

In Digicel's Jamaican market, the company responded to beeping and flashing by making it possible for users to request a call from other users by sending multiple free 'call me' texts (Horst and Miller 2005: 762). According to Horst and Miller (2005), this response was welcomed and recognised as a popular gesture acknowledging how Jamaicans communicate with each other. The service was introduced into PNG soon after Digicel's launch in July 2007 (see Chapter 2, Figure 2.7). By 2014, it was so popular that one man commented to me that only five-year-old children and very old people do not know how to send 'call me' requests. In effect, the service would permit one user to subsidise another user, ensuring that rent would be paid for use of the network. One could, however, respond to a 'call me' request with a 'call me' request of one's own, thereby politely indicating that one had no credit to make a call.

Digicel subsequently introduced an innovation on the 'call me' text request: the 'Credit Me/Credit U' service. This service allows a user to request a transfer of credit from another user's account by sending a text specifying the amount requested. The requested amount could be sent for a small fee, initially 20 toea but raised subsequently to 45 toea by 2018. This service not only enabled one user to subsidise the calls, texts and data of another user, ensuring that rent would be paid for use of the network, but also generated revenue for the company in the process. Users, however, already had a workaround in place that allowed them to avoid paying the fee for credit transfers. A user could transfer credit to another user by purchasing a flex card, scratching off the coating to reveal the code, then texting the numbered code to another user, who could add credit or top up an account balance by entering the number.[1] This transfer cost only the price of the text, which was usually as nominal as 2 toea for texts purchased in bundles. Moreover, it would enable the recipient who entered the number to receive any 'rewards' (such as bonus minutes; see below) being offered by Digicel in connection with topping up.

In April 2020, Digicel's Credit Me/Credit U service was still operating, but without charging any fees. The free service was advertised as 'one of the many ways to stay connected during Social Distancing' necessitated by COVID-19 lockdowns. Digicel's gesture, hardly free of self-interest, responded to the difficulties some users encountered in acquiring phone credit during lockdowns. By 2022, a service fee of 50 toea had been reinstated for Credit U transfers. Moreover, the number of 'credit me' requests were limited to seven per week.[2] This limitation was perhaps an attempt to repossess the 'credit me' service from users, that is, to exert some control over how users had developed an impressive repertoire of messages that could be sent using 'credit me' requests. This development

---

1   Two individuals told me that they saved used flex cards in the hope that the numbered codes would be recycled and therefore might work again in the future.

2   In Tonga, a Digicel subscriber in 2022 could make 'credit me' requests only when the subscriber's balance is 'less than 0.12 cents' (web.archive.org/web/20230206094636/support-to.digicelgroup.com/hc/en-us/articles/360033920751-How-to-send-Credit-U-Credit-Me, accessed 26 December 2022).

built upon the creativity expressed worldwide in formulating SMS lingos—ways of texting on basic phones adapted to the challenges of cramming information into a limited number of characters and dealing with small keypads that incentivised economical keystrokes. For example, 'SMS lingo makes extensive use of abbreviations, such as "C U l8er" for "See you later," "d8" for "date" or "B4" for "before," "communic8" for "communicate," etc.' (Temple et al. 2009: 1; see Temple 2016).

In PNG, credit requests were enlisted by users in service of sending messages other than 'please credit me'. This practice drew upon the large set of numerical abbreviations already in use to send messages, such as 99 for 'good night' ('nine nine' sounds like 'night night') and 43 for 'love you' (when written '43' resembles the image of a heart turned on its side). So, for example, a credit request for 99 kina would be understood as a friendly 'good night' by the recipient. Credit requests could also be used to reply to texts. A text that asks 'When will you be free to meet?' might be responded to with a credit request for 12 kina in order to indicate '12 noon'. Among students at the University of Papua New Guinea (UPNG), these requests followed local conventions to indicate a person's whereabouts. For example, the number seven referred to the library; a credit request for 71 kina indicated a location of 'library, first floor' (Temple et al. 2009). These conventions stretch the analogical imagination: the number 8 referred to 'eyeglasses', and hence a request for 8 kina indicated that a person was 'reading in the library'!

***

University students enthusiastically took up texting as an inexpensive alternative to voice calls and an expressive style of communication that was cool as well as opaque to the monitoring of uninitiated outsiders. Along with SMS lingo, other novel forms of usage in Tok Pisin—the creole language widespread in PNG—emerged around mobile phones. Handman (2013) discusses how the orthographic choices made in text messages, which often incorporate Anglicised Tok Pisin and Pidginised English, reflect a youthful sense of cosmopolitan modernity. King (2014: 125) records a number of innovations that enable Tok Pisin speakers to talk about their mobile phones, such as this borrowing: '*sampela taim mi blututim ol singsing mi laikim*/And sometimes I Bluetooth songs that I like'. (Bluetoothing, of course, also enables users to access music without having to pay for the data otherwise required to download it; see Crowdy and Horst 2022). Temple (2016) notes creative analogies based on mobile phone usage such as the expression *'lalu ful bar/*Love you full bar', which likens the intensity of love to a signal for strong network connection. All of these examples illustrate how the appropriation of mobile phones is caught up in literal and figurative processes of creolisation, generative processes of mixing and remixing that register in and through words such as *blututim* and actions such as saying good night with a 99 kina credit request (see Bar et al. 2016).

PNG urban residents—especially youth—also took to texting with gusto, sparking debate about whether text messaging was leading to a decline in proper language skills (grammar and spelling) or providing a compelling reason for young folks to learn how to read and write. Digicel clearly recognised the value of texting and was quick to offer its users discounts for repeated texting. In 2010, it was possible to send a text for the price of 25 toea. However, after sending three texts in a day, all subsequent texts would be charged at the nominal price of 1 toea each. A few years later, Digicel was offering its users a deal in which two texts would cost 50 toea, but the next 18 texts would be 'free'. This deal substantially reduced the per text cost for 20 texts from 4.6 toea to 2.5 toea.

By 2014, Digicel was offering a new deal for text messages: a bundle of 50 text messages for 75 toea, a further reduction in the per text price to 1.5 toea. This price reduction, however, did not please everyone. One man, a security driver for a Port Moresby cultural organisation, complained to me that 50 texts were too many to use in a single day. Indeed, it was not uncommon at that time to receive a spate of good night texts as the clock approached midnight when the text bundle would expire. Not long afterwards, Digicel introduced another new deal, one still available in 2020: 60 texts for 1.20 kina, or 2 toea per text, as was explicitly noted in advertisements for the deal. This deal actually raised the per text price by 25 per cent. But that was not the only reason the deal attracted criticism. One young man, a tutor at UPNG, told me that by raising the price to more than 1 kina, Digicel had made it impossible for him to send or receive 1 kina in order to purchase a text bundle. Exchanges of credit for this purpose were a routine feature of his interactions with his girlfriend (see below).

My point in sketching this brief history of text message deals is to demonstrate how over the course of a few years, the size of the discounted bundle grew 300 per cent from 20 to 60 messages. This growth illustrates how users were enticed by Digicel to increase their usage (if not necessarily their total expenditure). Indeed, consumers were regularly exposed to the company's inducements to consume more in the form of the promotions discussed in

Chapter 2 and in the form of 'rewards' or bonuses offered for regular usage. Rather than putting constraints on the freedom of users to communicate, the company effectively made users responsible for constraining themselves from excessive consumption. The example of how one controversial reward tested the self-discipline of users will illustrate the dilemma that users confronted.

First, however, it is worth emphasising how prepaid subscriptions entail a particular moral economy in which mobile users assume fiscal responsibility for managing their phone credits. Personal discipline is intrinsic to the model of prepayments; it is structural. The basic terms and conditions of prepaid subscriptions impose definite constraints. One is, of course, able to use *only* airtime that one has already paid for, but one also *must* use the airtime one has paid for within a certain period of time or else forfeit it.

Consider the following terms and conditions, which were in effect around 2015. Digicel PNG airtime credits would expire after 30 days, although they might be preserved by topping up one's balance before the expiration date. Likewise, subscriber identity modules (SIM cards) would expire when they have not been used for a certain period of time that might range up to one full year. In PNG, Digicel SIM cards expired after three months. Sale of data credits for use on laptop computers or mobile phones was also organised in terms of expiration dates. In PNG in 2015, one could buy data credits from Digicel that could be used for one hour (10 megabytes [MB] for 99 toea) or for 30 days (1,500 MB for 60 kina) with several options in between.[3] Each option, however, came with the caveat of 'use it or lose it'. That is, prepayment applies a temporal discipline upon subscribers, one that strongly encourages if not actually enforces consumption. A colleague of mine accordingly confessed that she would stay awake streaming videos on her laptop when she knew that her data credits were about to expire, despite the fact that she had no real interest in the videos and always regretted her decision the following morning.

Second, discipline is subjective, in all senses of the term. Prepaid subscriptions devolve responsibility for managing credits upon the individual using the phone; they encourage the formation of what Donner (2015: 123) calls

---

3   The price of data has dropped considerably since 2015. In 2023, Digicel data plans ranged from 100 MB for 24 hours for 3 kina to 80 gigabytes (GB) for 600 kina for 30 days, with many options in between (support-pg.digicelgroup.com/hc/en-us/articles/7313007383439-What-are-the-new-data-plans-, accessed 6 July 2023, site since discontinued). In 2023, bmobile data plans ranged from 500 MB for 24 hours for 3 kina to 30 GB for 30 days for 150 kina with many options in between (www.bmobile. com.pg/data, accessed 25 May 2023).

a 'metered mindset'. One must regularly check the balance of one's account; be aware of the length of calls and whether they are being charged at peak or off-peak rates, or at on- or off-network rates; and be aware of when precisely a data bundle will expire—after which time one's phone credits will be charged at very steep 'out of bundle' rates (Digicel charged 49 toea per MB in 2015).[4] It is up to the individual, not the company, to ensure that he or she has the capacity to make a call or to go online. And that includes, by the way, making sure that the phone's battery is charged, a condition never taken for granted by the vast majority of Papua New Guineans who have uncertain and limited access to electricity (see Chapter 1).

Mobile phones, in other words, are devices for promoting a sense of time thrift, an overriding consciousness of the fact that time is money. Linda, a woman from Bougainville living in Eastern Highlands Province in 2015, told me that Bougainvilleans speak slowly: 'Helllooooo. How are you dooooing?', she mimicked.[5] She then admitted that, whenever speaking with her family in the village, she is constantly thinking 'Hurry up! Talk quickly! This is costing me!' It is in this regard, I suggest, that mobile phones are effective instruments for teaching an apprehension of time that would have pleased Benjamin Franklin (see Weber 1992)—certainly more effective than the discipline of industrial factory labour, which is in any case extremely rare in PNG. (It is hardly surprising that Digicel was able to create goodwill in many of the Caribbean markets it first entered by offering per-second billing for calls instead of charging by rounding up to the nearest 30 seconds or minute [Horst and Miller 2006].) Moreover, the sort of personal discipline that mobile phones encourage recalls other technologies—such as self-help programs (Bainton 2010, 2011) and fast-money schemes (Cox 2011, 2018)—that Papua New Guineans have embraced in search of financial viability and moral legitimacy.

The dilemma of self-imposed constraint in the face of an opportunity to consume freely surfaced in one particular Digicel promotion that was running in 2016. The promotion rewarded users who top up their balances in the amount of 5 kina or more. A user would receive from Digicel the following text message: 'Congratulations! You have been rewarded with

---

4   Another way to manage the fiscal exigencies of prepaid subscriptions is to acquire a dual SIM phone that allows a user to purchase a cheap data bundle from one company and unlimited calls and texts from a different company.
5   I use a pseudonym here and elsewhere in the book when providing ethnographic examples of mobile phone use.

100min talk time bundle. 100 mins valid for local mobile calls between 11PM–7AM.' These minutes could only be used beginning 11 pm on the day that one tops-up; the bundle would expire at 7 am the next morning.

This 'free nights' promotion provoked strange mixtures of restraint and excess that confound any easy answer to the question of what sort of fiscal subjects are constructed by prepaid subscriptions. The notion of time thrift—at least as Ben Franklin would recognise it—seems perverted when people stay awake until the wee hours of the morning to make phone calls. I recall one sardonic letter to a newspaper asking whether Digicel thought the author was a flying fox, expecting him to be awake for conversations at 3 am. Indeed, this aspect of the promotion received public criticism from parents complaining that their children were staying up too late on school nights. Comments posted to Digicel's Facebook page in 2015 show how the concerns raised by the 'free nights' promotion were moral concerns that put in question the very legitimacy of the relationship between Digicel and its subscribers. For example, one commenter wrote:

> As a concerned parent this particular promotion encourages a lot of parental/youth related issues. It should be considered. How much this contributes to many staying up late and early hours chatting due to such promotions. Please have some consideration for family and peer issues in this country. The effects on society and the growing and future generations and education. In Business you make your money but be responsible also. Thank you.

On the other hand, time thrift and calculative agency were undeniably present. Users deferred making calls during peak-rate periods in order to take advantage of the promotion. One tired young man reported to me that he had received a call from a friend at 1 am. When he anxiously asked the friend if anything was wrong, the friend explained that all was well, but he just did not want to 'waste' the free minutes he received through the promotion. Likewise, commenters on Digicel's Facebook page complained that because they were asleep by 11 pm their 'free minutes' were wasted. This sort of time thrift risked provoking a negative response for reasons besides disturbing people's sleep. One young woman thus observed that she would not be pleased if her boyfriend called her with free minutes, since it would imply that she was not worth the call otherwise. The currency of phone credits here measures the value of intimacy.[6]

---

6   This promotion also stimulated the annoying practice of making random calls, sometimes repeatedly to the same number, in search of cross-sex 'phone friends' (Andersen 2013; Jorgensen 2014; see Chapter 5).

One logical response to Digicel's promotions and promptings—to its efforts to stimulate more consumption of voice, SMS and data—is greater self-imposed discipline. Let us consider again the example of Linda from 2015, not as typical of Papua New Guineans, but as illustrative of the way in which autonomy and dependence, self-discipline and self-indulgence, freedom and constraint all merge in the everyday micro-practices of operating a mobile phone. Linda is a heavy user who can spend up to 100 kina every week in phone credits. She is a young single woman living far from home and regards frequent communication with her family and friends as nothing less than essential. Linda, who has a steady income, is aware that she is capable of spending all her savings on airtime, and she has on occasion come close to doing so by topping up her phone through a mobile banking account with Bank South Pacific—a relatively new service in 2015 that effectively enabled users to top up anytime, anywhere. (A suggestive comparison might be made here with gambling machines that provide access to a player's bank account, thereby enabling the player to remain seated in front of the machine while withdrawing funds to continue playing.) In order to discipline herself—and Linda used the English word discipline—Linda opened an account with another bank into which she makes weekly deposits. This bank, according to Linda, did not offer the mobile top-up service. She thus safeguarded her money from herself.

Jack, a Telikom worker, suggested to me that the sale of flex cards in small denominations promotes waste on the part of people who cannot afford such purchases. Jack claimed hyperbolically that people buy 3 kina flex cards 10 times a day without realising that they have spent 30 kina. Instead, he proposed, the minimum price for a flex card should be 10 or 20 kina. That way people will feel the pinch at the point of purchase and only people who can really afford it will buy the flex card.

Linda, in contrast to Jack's imagined careless consumers, was explicit about the calculations that she made in managing her mobile phone. Linda claimed that she did not feel able to start the day unless she is equipped to communicate. So, in the morning she would top up her phone for 5 kina. This top-up gave her the aforementioned 100 free promotional minutes to use that night starting at 11 pm. She then purchased a one-day data pass—60 MB for 3 kina, at the time. This data was enough to allow her to go online and communicate with friends and family via the applications WhatsApp and Viber. Linda had only recently discovered that she could send voice messages over the internet for much less money than making voice calls. Finally, Linda purchased 60 text messages for PGK1.20.

She would use most if not all of these text messages before they expired at midnight. That left 80 toea as a balance in case Linda needs to make a quick phone call during the day. (On-net calls from one Digicel phone to another were at the time billed at 79 toea per minute during the peak hours of 7 am and 9 pm.) Once she had made these preparations, Linda felt ready to go out into the world and meet the demands of the day.

***

A further turn in the technology evolution cycle occurred in 2015, when Digicel introduced a scheme whereby users could borrow credit in advance. Users low on credit would receive a text message offering them 3 kina of credit with the provision that upon the user's next top-up, 3.90 kina would be deducted; in other words, a loan with 30 per cent interest.

In 2020, the scheme was called 'Credit and Bundle on Loan' because it offered the option to use the loan to buy bundles of voice calls, texts and data as well as airtime credit. If the Credit Me/Credit U and Call Me services allowed a user's friends and relatives to subsidise usage, the Credit and Bundle on Loan service allowed a user's future self to subsidise the user's present self.

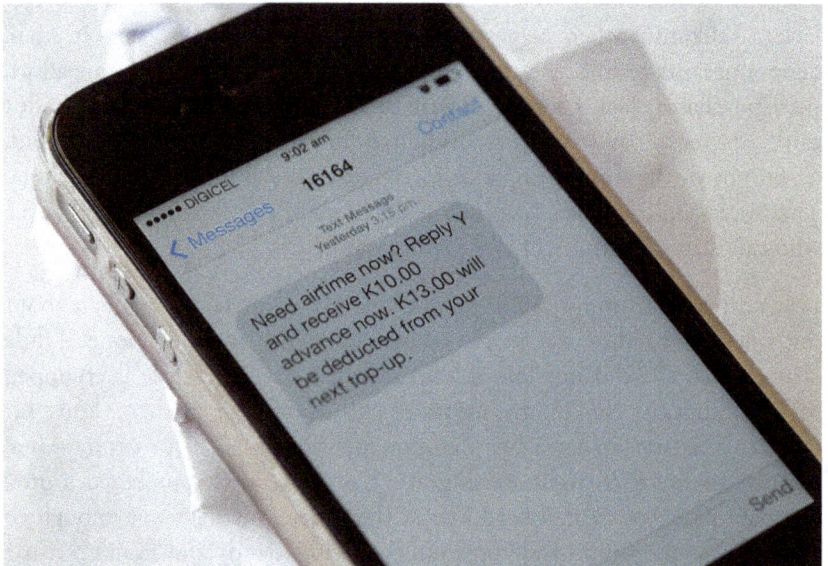

Figure 3.2. Digicel text message for a credit loan.
Source: Photo by W Bai Magea.

Digicel's loan service was instantly popular. In 2015, a University of Goroka student admitted that 'most young people I know owe Digicel 3 kina'. Another man pointed out that the service was very convenient if one ran out of credit at night. Instead of searching for someone from whom to purchase a flex card, which would probably have a fee of at least 50 toea added to its face value, one could take out the loan and top up safe at home at any hour, using either a basic ('one-bang') phone or a smartphone. In this sense, the loan service was—along with emerging options for users to top up online through their linked bank accounts—another blow to folks who sold flex cards and electronic top-up for a living (see below). Rooney et al. (2020) have rightly asked if Digicel's loan service charges predatory interest rates and whether it is part of a larger microlending environment that exploits vulnerable urban dwellers. They also raise the question of whether, as in the case of gambling, Digicel as a mobile network operator rather than a financial institution is licensed to offer loans, thus calling attention to the moral obligation of state agencies such as the Independent Consumer and Competition Commission (ICCC) and National Information and Communications Technology Authority (NICTA) to protect the welfare of consumers, including Digicel subscribers.

I would add to these important questions the observation that Digicel's loan service illustrates well the power dynamics of appropriation. Digicel for its part had to manage the risk that many of its loans would go unpaid, thus enabling debtors to use the network rent-free. The company managed this risk by restricting loans to subscribers who meet certain criteria based on their recent usage, a record of which is available to Digicel. According to Digicel's website, subscribers must have used their SIM cards for at least 30 days and topped up in the last seven days.[7] Moreover, the size of the loans offered varies depending upon how much a user has topped up in the previous four weeks; users who have topped up by more than 100 kina are eligible for larger loans than users who have topped up between 10 kina (the minimum amount) and 40 kina. Finally, if a loan is used to purchase a bundle—say, an 8 kina loan for 80 on-net minutes, 80 SMS, and 120 MB of data—then the credits will be valid for one day only: a strong incentive to use and not lose the credits.

---

7   See: support-pg.digicelgroup.com/hc/en-us/sections/6454539658639-Credit-and-Bundle-on-Loan, accessed 6 March 2021, site now discontinued.

Users, for their part, experimented with ways to work around the safeguards that Digicel applied against using the network rent-free. The option of taking out a loan and then never using the SIM card again was of course available, but perhaps not economical, especially given the time and effort required to register SIM cards since 2019 (see Chapter 5). But other options were explored. The University of Goroka student who noted the popularity of 3 kina loans in 2015 also noted that her Facebook friends discussed ways to purchase data without having the 3.90 kina deducted from a debtor's newly topped up account. (Users can also visit online forums such as the Digicel Complaints Group Facebook page to find tips and share advice about workarounds.) For example, the student suggested that a user could avoid repaying a loan by purchasing a data bundle before the interest due was deducted: 'type "DAY" to 1634 on standby. As soon as he [the user] recharges his account he must press send'. This quick action would allegedly result in the purchase of a day pass for 60 MB data before the deduction for the outstanding debt could be made. Similarly, one user claimed that even though he had an outstanding debt, he could receive credit sent through the Credit Me/Credit U service without having to repay the loan—a brilliant instance (true or not) of appropriating the company's repossession of an earlier appropriation. The negotiation of power relationships embedded in the technology of mobile phones, although never quite symmetrical, is also never completely foreclosed.

Gift Economy

In addition to indicating the negotiation of control over a new technology, the notion of appropriation can be taken to mean the appropriateness of the technology to its new social and cultural context. To what extent does the technology afford practices that resonate with prevailing cultural values and fit with entrenched social conventions? To what extent does it afford practices that present challenges—whether subtle or obvious—to these values and conventions?

These questions have caught the attention of scholars studying the use and effects of mobile phones throughout the Global South. Anthropologists, in particular, have noted how new forms of privacy enabled by mobile telephony have disrupted and reshaped social relations between seniors and juniors, husbands and wives, mothers and daughters-in-law as well as between peers and friends (see Chapter 5). Almost all of these cases demonstrate how the uptake of mobile phones can both reinforce old habits and unleash potentially transformative practices with respect to gender relations and personal autonomy—leading to a profound sense of ambivalence on the part of all concerned.

Predictably, then, anthropologists working in Oceanic societies have sought to understand how mobile phones fit with the practices of gift exchange and norms of reciprocity associated with the region as 'gatekeeping concepts' (Appadurai 1986) since the publication of Malinowski's 1922 account of *kula* and Mauss's celebrated *Essai sur le Don* (1923–24). Some of their findings, perhaps surprising to casual observers, have been reported with a dash of sensationalism in popular media. For example, *The New Republic* (Robb 2014) reported on research by Telban and Vávrová (2014) that discussed how Karawari people in the East Sepik Province of PNG were using mobile phones to call the dead. The practice of communicating with deceased relatives was not new; only the means to communicate were new. Nor was the aim of the communication new: namely, to ask the dead to provide knowledge and goods, and to put money in the bank accounts of the living. In other words, the mobile phones were put to hopefully more effective use in activating a gift relationship between the living and the dead: 'From this perspective, the anticipated access to mobile connection and communication represents a means to achieve something that has always been there, although persistently concealed and never fully accessible' (Telban and Vávrová 2014: 2–3).

David Lipset's (2013, 2018) ethnography of mobile use by people of the remote Murik Lakes area, near the mouth of the Sepik River, demonstrates the possibility of subsuming the new technology within local understandings of the moral obligations that kin and neighbours ought to bear toward each other. Lipset argues that Murik villagers, as well as Murik migrants in Wewak town, use their mobile phones in ways that caution against the assumption that the technology is somehow inherently individualising and individualistic. For example, in Darapap village, where network reception is unreliable and uneven, a handset hangs from a cord in a doorway of the home of a young couple. This handset occasionally gets reception in the early mornings and at night. The young couple allows fellow villagers to use the phone and takes incoming calls, often from Murik in town, and delivers messages to the intended recipient. These calls almost always concern the wellbeing of kin or the long-distance coordination of kinship-related activities such as funerals. There is no fee for this service, but rather an open-ended expectation that phone users will now and then provide the couple with credits. Hardly a token of the modern possessive and private individual, or urban 'big shot' (see Chapter 1), the handset is instead a sign and instrument of a moral community underpinned by a perceptible gift economy.

Phone sharing of the sort described by Lipset for Darapap village appears to be less common in Port Moresby and Goroka, where I rarely encountered it among the scores of people interviewed and surveyed for this research. Even school-age children in Port Moresby would have their own phones, often provided by their parents as a measure of security in the uncertain environment of the city. But if phones themselves are rarely shared, they are commonly used to facilitate other kinds of sharing. One man recalled how in the early days of mobiles even the simplest Nokia phones had a ring-tone composer that could be used to create polyphonic ringtones that were sent to others via SMS. Music still circulates from phone to phone via Bluetooth, which is recognised as free but very slow. More recently, music has begun to be shared by members of WhatsApp groups, and through wi-fi sharing programs such as CShare and ShareIt (Watson, Crowdy et al. 2020; Crowdy and Horst 2022). Music videos (many of which feature songs about the travails of using mobile phones) and YouTube videos are likewise shared: 'An application called Tubemate was commonly used to save video content from YouTube locally, thus avoiding to have to download it more than once and saving data' (Watson, Crowdy et al. 2020). Mobile phones and Bluetooth are also used, I was told, to share pornography, the circulation of which in PNG caused loud public concern following a Google Trends report that the country ranked first in the world for internet 'porn' searches (Watson 2017). (See Hobbis [2020] for an extensive discussion of how smartphones are used for sharing music, videos and photographs in Solomon Islands.)

Generalised or open-ended sharing is something different from gift exchange that follows definite protocols and prescribed channels. The ways in which mobile phone users make and respond to 'call me' and credit requests can upset or redefine these protocols and channels. Good (2012: 212) attributes the popularity of Digicel's Credit Me/Credit U service in Tonga to its resonance with the cultural values of loving generosity and, specifically, the practice of *kole* ('to beg, to make a request'):

> It was expected that if a *kole* request was made, the addressee had a moral obligation to fulfill that request if she had the resources to do so, even if that meant putting herself at a disadvantage. (Good 2012: 213; see Watt [2019] and Peseckas [2014] on mobile phones and similar requests—*kerekere*—in Fiji)

But *kole* requests could not be made to just anyone. Among the Tongan youth that Good wrote about, it was usual for a young man to satisfy the *kole* requests of a young woman, for example, by supplying the young woman with phone credits to use for communication with him. However, young women began to transfer phone credits to young men, reversing the traditional direction of exchange and renegotiating both control of the social relationship and the gendered expectations of proper gift exchange.

Generalised or open-ended sharing is also something different from calculated giving and receiving, and the exacerbated tensions between the two are a feature of the encounter across Oceania with the logics of commodity exchange. In Vanuatu, young men born and raised in the capital Port Vila regard mobile phones as crucial tools for building social networks that extend and enhance status and reputation (Kraemer 2015, 2018). These young men, however, often lack the financial resources—that is, credit—to use their phones. Accordingly, they cultivate relationships with others to whom they can make requests when necessary and discourage relationships with others who are thought only to make requests of them. There is a definite sense of calculative rationality involved in 'working the mobile': 'Giving and spending phone credit in a relationship has become a signifier of the value of that relationship to a person' (Kraemer 2018: 102). While it would be wrong to say that such calculations are not a part of traditional gift exchanges, it is fair to say that mobile phone credits have become a new currency in which to measure, modulate and maintain 'a desired degree of intimacy' in social relations (Kraemer 2018: 102).

The ethnographic evidence from PNG confirms and broadens the findings about credit requests reported from Tonga and Vanuatu. Research, including a review of two dozen diaries documenting the phone use of university students over a 48-hour period in 2014, suggests that almost everyone uses this service. Regular credit exchanges of small amounts (usually 1 kina) were a common feature of boyfriend/girlfriend relationships among the students surveyed, who also requested larger amounts of credit from their parents and older siblings with similar frequency. In the years before smartphones became common, transfers of 1 kina were often used to purchase 50 SMS for 75 toea as a conversational alternative to more expensive voice calls. Hence the resentment among many users when Digicel raised the price of the text bundle to 1.20 kina for 60 SMS, thus requiring a transfer of 2 kina to cover the cost.

The exchange of texts and credits between boyfriends and girlfriends, according to data from the diaries and interviews with students, entails a set of tacit expectations. For example, there is an expectation that a text (or call) should receive an immediate reply. Texts that go unread because the battery in one's phone needs a recharge can generate anxiety and worry. Similarly, network breakdowns that interrupt the flow of communications cause consternation. A frustrated poster on the Digicel Complaints Group Facebook page said:

> This bloody network is causing inconvenience for Family & Loved Ones … All the messages from Sunday just coming in now … Fix the problem coz if my GF [girlfriend] breaks up with me tomorrow than I will hold you responsible. (see Foster 2020)

This example highlights not only an unfortunate intersection between non-human infrastructure and human intimacy, but also the mistrust and suspicion that complicate mobile communication among youth, in PNG as elsewhere (Archambault 2013; Gilbert 2016).

The exchange of texts and credits, like the exchange of gifts broadly characteristic of Oceanic societies, incurs the obligation to uphold the relationship that the exchange brings into being and reproduces. Consider the following paraphrased story related by one of the university students who assisted with our research on mobile phone use in PNG:

> A friend of mine credited me K2 and I called him back to say thank you. He told me to get free SMS so we can text. I ended the call and subscribed for 60 free SMS by sending 'SMS' to +1629. As soon as I received notification from Digicel, I started texting with him. We started texting jokes and poking fun at each other. Then he said 'Uhm Winnie, before we run out of free SMS, I have to tell you something.' He told me how his girlfriend is seeing someone else and he liked her a lot but how can he tell her that. I told him not to do anything. Just get his mind busy with other things. I suggested that he start dating other girls just to get over her. We kept going on and on until I received a text message telling me that I have used up all my free SMS. I was happy. It was a relief for me. Finally, I would get some sleep. He kept texting and I sent him a 'Please Call Me' request as code for saying I have no credits. He sent me another K2 …

Winnie met her obligation to receive the additional gift and the conversation continued into the night even though Winnie's responses were often one or two words. When Winnie told this story to her girlfriends, they shared stories about how male friends had engaged them in similar ways. One of Winnie's friends commented wryly that at least unlike her, Winnie did not have to supply the credits.

The burden of receiving gifts of credit and dealing with the consequences that such gifts engender is forcefully demonstrated by stories of 'phone friends' reported by Barbara Andersen (2013; see Chapter 5). In one story, a husband and wife conspire to extract credits from an unknown man who calls the phone they share (sharing a phone in this story indicates the couple's status as rustic village people). The husband encourages his wife to entertain the caller, who begins to send money to the husband's bank account. Eventually, the unknown man reveals his plans to visit the woman, who had previously told the caller where she lives (another sign of her rural naiveté). When the man arrives, the husband pretends to be his wife's brother. The unknown caller, however, declares his plans to take the woman away as his wife. Her husband is trapped by the situation he has created: 'by consuming the money sent to his wife, the husband had entered a simulated—but still effective—affinal relationship' (Andersen 2013: 330). The moral of the story is to beware of accepting gifts from unknown sources, because even gifts received under false pretences carry obligations. Women in particular who accept such gifts risk becoming detached from their social moorings and lured away by the influence of 'subversive transactors' (Andersen 2013: 330).

Managing the back and forth of credit transfers and 'call me' requests usually involves interacting with relatives and friends rather than unknown callers and anonymous texters. Such is the conclusion of interviews conducted with a variety of people living in and around the town of Goroka in 2015. In this regard, it is important to note that 'call me' and 'credit me' requests are not always met. On the one hand, handling such requests is a matter of self-discipline. Some of the people interviewed reported restricting their requests to a few close family members; others claimed to find it hard to make requests because of pride. One interviewee similarly stated that he does not answer any of the requests that he receives because if a person has a mobile phone, then that person should be able to afford to operate it. Here, apparently, is evidence of a discourse of self-responsibility taking shape around mobile phones—exactly the sort of discourse that scholars who associate mobile phones with emergent forms of modern possessive individualism might expect. At the other extreme, however, Linda, the woman from Bougainville, claimed that she responds to all requests and even sends credits unsolicited to friends and family members. Linda explained that she was the firstborn in her family and had numerous siblings and other relatives to look out for.

On the other hand, handling requests is a matter of strategy. Indeed, the low balances that people keep on their phones can plausibly be interpreted as a strategy for protection against credit requests—a strategy that recalls traditional practices for safeguarding resources (betel nut, tobacco, garden produce, shell valuables) from the claims of others by keeping them out of sight and inaccessible (Foster 1993). Crowdy and Horst (2022) claim that several of their research participants likewise deliberately choose not to carry on their phones music that they do not wish to share. In many instances, people meet requests selectively, deciding when it is appropriate to respond positively and to whom. For example, one man living in Port Moresby, Basil, explained that when Digicel installed a tower in his home village, he changed SIM cards so that relatives at home would not pester him with 'call me' requests that would lead to 'unnecessary' calls. When asked what he meant by unnecessary, Basil gave the example of someone telling him that a pig broke into a neighbour's garden: 'This is none of my business. Why waste my time?' Even Basil's elder brother in the village has to call Basil's wife, who then communicates a message to Basil if it is deemed necessary.

One older woman living in Goroka explained that she responds to her children's 'call me' requests, but not to their 'credit me' requests. That is, she is willing to pay to speak with her children, but not to finance their communication with other people. This sentiment also informs gifts made between boyfriends and girlfriends, which are presumed to be used exclusively as a resource for the couple's private communications—a presumption not always justified (see Good 2012: 215). The sentiment might count as evidence of what critics of neoliberalism understand as the extension of economic rationality into the domain of the social (see e.g. von Schnitzler 2008). Like displeasure at being phoned with so-called free minutes at off-peak hours instead of during the evening at peak rates, the woman's denial of her children's credit requests registers the calculation of relative intimacy in terms of the price of airtime.

Informal Economy

The spread of mobile phones across the Global South was accompanied by expectations that an era of robust economic development was at hand. Mobile phones, it was thought, would enable new forms of entrepreneurship and provide women, in particular, with new means for achieving economic autonomy (see Curry et al. 2016). In PNG, there is some evidence of the positive impact of mobile phones on the growth of businesses, especially the growth of taxi services in the capital Port Moresby. Similarly, there is also clear evidence of the positive impact of mobile phones on organising commercial transport—in the long-distance betel nut trade, for example (Sharp 2012: 122).

Facebook has made the business of at least one Port Moresby fish monger much more efficient. The vendor posted early morning photos of fish for sale on her Facebook page with a range of prices, contact information and the promise of delivery at no extra cost. This service reportedly appealed to city residents who were imagined to be either 'too lazy' to visit the marketplace in person or who were hesitant to venture into locations known to be frequented by pickpockets and purse snatchers.

Anthropologists, however, have been quick to temper inflated expectations of the economic impact of mobile phones. Ethnographic accounts have curbed the enthusiasm with which some advocates of mobile phones promote their entrepreneurial use in the developing world. The work of Horst and Miller (2006) in Jamaica, for example, demonstrated that the promise of economic freedom attached to mobile phones by development agencies was rarely redeemable. Mobile phones did not stimulate the sort of entrepreneurial activity that was expected, although they did indeed enable poor people to cope better with their circumstances. Paradoxically, as Miller notes (2006), the latter was a corollary of the former. That is, mobile phones enabled people to appeal to each other more effectively in order to share resources when emergencies arise and provided easier access to friends and family members living abroad, who might have resources available (Horst and Miller 2006; Horst 2006). But it was precisely this sharing of resources that prevented the formation of capital needed to start new businesses and to flourish as a self-made entrepreneur.

When it comes to starting or expanding businesses, mobile phones often favour people who already have the resources to operate successfully (see e.g. Wallis 2011). Recent ethnographic research on the economic impact of mobile phones on coffee farming and fresh produce trading in the eastern highlands of PNG likewise reports mixed findings (Titus 2019). While mobile phones certainly enabled better coordination of transport, little evidence was found that mobile phones led to 'opportunities for arbitrage and better market prices for smallholders' (Titus 2019: 139; but see Suwamaru [2015] for claims about producers using mobile phones to secure better deals). My own inquiries in 2015 confirmed that there was little uptake on the part of coffee farmers and traders in the use of mobile phones for the mobile money scheme (Cellmoni) that Digicel was promoting at the time. By contrast, there is strong evidence of the use of mobile phones, especially in urban areas, for banking—transferring funds, checking balances, paying wages, purchasing electric power and, of course, topping up phone credits (Suwamaru 2015).

What both proponents and sceptics of the economic power of mobile phones tend to underemphasise is the informal economy that advances and results from the spread of mobile phones (but see GSMA 2021). This fact was not lost on Digicel, which occasionally claimed credit for the phenomenon in PNG. Thus in 2019, Digicel Senior Vice President Lorna McPherson, in comments defending the company's contributions to PNG, asserted that:

> We have over 800 staff, but also have around 35,000 people that resell our products and services. So it's not only Digicel's own employees but it impacts the economy and the way we do things is huge. (Post-Courier 2019)

Soon after launching, Digicel publicised the good it was doing in PNG in the figure of the street vendor (*strit venda* in Tok Pisin) who resells phone credits in small amounts to retail customers.

Consider the newspaper advertisement that ran on 2 October 2007 in *The National* under the heading 'My Life Is Better!'

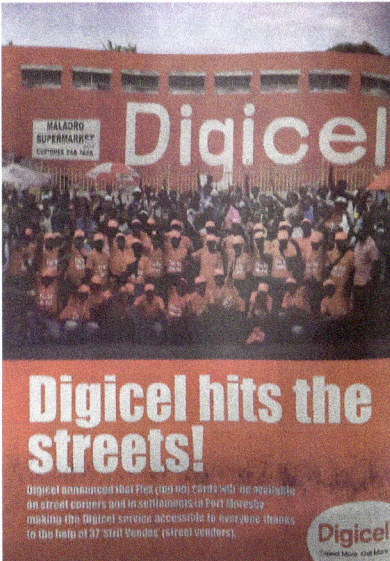

Figure 3.3. Digicel newspaper ad (Street Vendor Programme).

Source: *The National*, 24 August 2007.

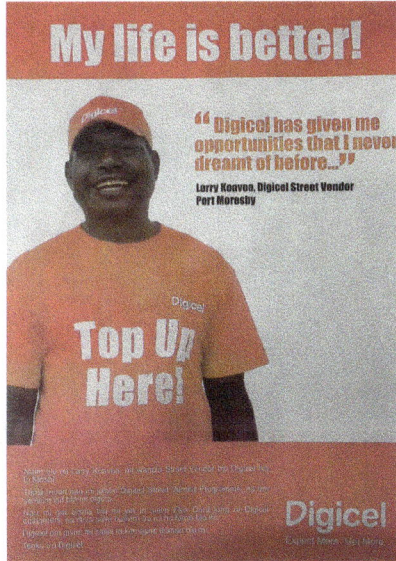

Figure 3.4. Digicel newspaper ad (Larry Koavea, My life is better).

Source: *The National*, 2 October 2007.

The ad featured an image of a young man, Larry Koavea, wearing a red baseball cap stamped in white with the name Digicel and a red T-shirt stamped with the words 'Digicel. Top Up Here!' The main text of the ad is a testimony, written in Tok Pisin and translated into English by me as follows:

> My name is Larry Koavea, and I am a street vendor for Digicel here in Port Moresby. Three months ago I joined the Digicel Street Vendor Programme, and it changed my life completely. Now I have my own business selling Flex Cards to Digicel customers, and this is really helping me and my family. Digicel gave me the chance to realise my dreams. Thank you very much, Digicel.

The ad was one in a series of testimonies to how Digicel had improved people's lives in PNG, including a Port Moresby bus driver who could call for help with roadside emergencies and a netbag (*bilum*) seller in Goroka who could stay in touch with her rural relatives in the Bena district of the eastern highlands.

Figure 3.5. Top-up transaction, Goroka, 2015.
Source: Photo by W Bai Magea.

In the first years of business, Digicel street vendors were instrumental in making the airtime credit necessary to make calls and send texts available for purchase beyond retail stores. Street vendors sold flex cards in urban settlements and at city bus stops as well as in the streets and marketplaces of provincial towns visited daily by rural residents for shopping, banking, selling produce and seeking health care. Vendors also sold electronic top-up: credit that they transferred from their mobile phones directly to the phones of customers. Some vendors offered basic handsets and memory cards loaded with music. SIM cards, with or without a handset, were sold on the streets prior to the implementation of SIM card registration (see Chapter 5). Digicel vendors were a common sight around Port Moresby and other urban centres circa 2015, often sitting or standing under red umbrellas sporting the Digicel logo. It was less common to spot a vendor sitting or standing under a purple and orange bmobile umbrella. Some vendors, however, sold both Digicel and bmobile flex cards.

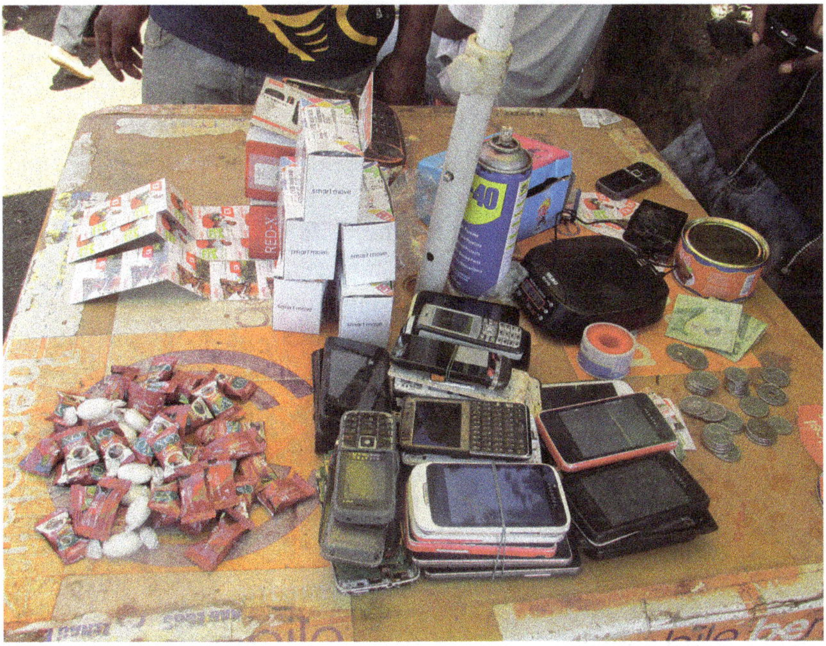

Figure 3.6. Mobile phone repair technician's table, Goroka, 2015.
Source: Photo by W Bai Magea.

As part of our research project, we studied and documented the activities of street vendors operating in the eastern highlands town of Goroka during 2015–16 (see Bai Magea 2019; Thomas et al. 2018). Street vendors were the most prominent actors in an informal economy that also provided repair services, battery recharging and phone unlocking (for forgotten passwords) to phone users. In Goroka, all of the street vendors selling mobile phone credits and basic handsets that we observed and interviewed were men, ranging in age from early twenties to mid-fifties (see Little 2016; Barnett-Naghshineh 2019). In other parts of PNG, such as the Rabaul/Kokopo area of East New Britain Province, women sell flex cards and electronic top-up at marketplaces and other public locations.

In Goroka, street vendors fell into two categories. Some street vendors worked at an interface between the formal and informal economy, buying flex cards and electronic top-up at discount from Oceanic Communications, a company contracted by Digicel to handle this aspect of the business, and then reselling to customers at retail prices, sometimes adding an additional service fee. These street vendors were individual businessmen who made their own hours, selling at a customary location for which they paid the

Goroka town authority an annual fee of 150 kina. Other street vendors worked as employees of one individual (a kinsman or *wantok* or fellow church member) who supplied them with flex cards and electronic top-up and who paid the town authority fees for the various public locations at which the employees set up their stalls. Stalls consisted of a small table, a chair and a large umbrella, often adorned with the logos of either Digicel or bmobile. (Vendors claimed that Digicel used to provide this equipment to street vendors for no charge; in 2016, however, tables and umbrellas were purchased for a fee of 200 kina.) These employees were paid a fortnightly salary or otherwise were compensated in kind with food and lodging. We learned of five groups operating in Goroka, each consisting of between five and 10 street vendors, including one group run by an individual who rented his various tables from another individual.

I limit myself here to reflections on the informal economy that align with the expanded idea of appropriation that informs this chapter. First, I ask: How do street vendors create ways to augment their income beyond the small margins gained through sale of discounted flex cards and electronic top-up? In 2015, vendors bought flex cards from Oceanic Communications at a discount. Flex cards with a face value of 3 kina cost 2.80 kina, and 5 kina cards cost 4.60. These two denominations were the ones most commonly sold in Goroka, although a few vendors sold 10 kina cards. Similarly, vendors purchased credit for electronic or phone-to-phone top-up at the rate of 107 kina worth of credit for 100 kina (to resell for a profit of 7 per cent; this rate had previously been as high as 10 per cent, according to one of the individuals who employed a group of vendors; Bai Magea 2019). Many street vendors complained that due to the proliferation of competitors, it was extremely difficult to make much money given the small margins granted by Oceanic Communications. One vendor remembered when he used to take in almost 200 kina a day in profits, up to 300 kina during the annual Goroka cultural show. Now, however, he struggled to make 50 kina a day in profits.

Vendors can augment their revenue by adding a service fee to each flex card. This strategy is less viable in a setting where there are numerous vendors like the streets of Goroka during weekdays, where the majority of vendors waive such a fee. However, individuals who buy cards in town for resale in villages where access to phone credit is limited invariably add a markup of 50 toea or more, as do urban vendors who sell cards from their homes at night or enterprising university students who sell cards in their dormitories. Similarly, street vendors added a 20 toea service fee to phone-to-phone top-ups of less than 3 kina (the smallest denomination of Digicel flex cards). Three kina is the standard unit for flex card transactions; when a buyer approaches a vendor, the unspoken assumption is that the buyer wishes to purchase a 3 kina card. Thus, it is resource-poor users who can only afford a 1 or 2 kina top-up (say, to buy a bundle of SMS) that regularly incur an additional fee.

A more creative but less transparent way in which street vendors overcome the constraints on growing income for themselves involves participation in the grey market for phone credits. Vendors can purchase Digicel credit from users at a deep discount and transfer it using the Credit Me/Credit U service. For example, Bai Magea (2019) reports that women who receive gifts of phone credit from boyfriends or 'phone friends' (see Chapter 5) sometimes sell it to vendors—one way in which women can convert gifts into commodities and so participate in the informal economy of mobile phones. Ten kina worth of credit will be purchased for 5 kina in cash; 100 kina worth of credit will be purchased for 60 kina in cash. These transactions are also attractive to users who receive gifts of top-up from their relatives living abroad. Digicel offers promotions that double or triple the amount of top-up purchased when paid for in the local currencies of relatives living in Australia or Aotearoa New Zealand (see Foster 2018: 114–15). This bonus can be converted into cash by recipients in PNG who sell it to street vendors. Vendors thereby deftly appropriate one of the company's own promotions to work around its inadequate compensation for street sales, while users subversively reverse the process by which the company converts cash, including foreign exchange, into its restricted currency of credit for Digicel products (voice, SMS and data).

Street vendors who resell credit acquired by these unofficial means must transfer it through the Credit Me/Credit U service and pay the appropriate fee. One user complained to us about receiving credit from vendors in this fashion because this sort of top-up does not qualify the recipient for Digicel's popular promotions, such as the 100 free night minutes of voice calls awarded with a 5 kina top-up. Other users, however, were said to prefer buying top-up in this fashion when they owed a debt for a credit advance because the amount owed would not automatically be deducted from their account.

Street vendors, as mentioned, offer not only electronic top-up and flex cards but also repair services. (Battery charging in Goroka was available at a few of the mostly Asian-owned trade stores in town, usually for a fee of

2 kina per recharge.) Two phone repair technicians who were interviewed in Goroka had some formal training in electronics repair but largely taught themselves how to mend mobile phones. Their services offer users whose phones suffer cracked screens, broken microphones and other injuries a less expensive alternative to buying a new phone (see Thomas et al. 2018). Users sometimes pay for these services by offering broken phones that repair technicians can either fix and resell or cannibalise for spare parts. Similarly, repair technicians sometimes purchase phones to harvest spare parts or to unlock and resell.[8] Some of these devices are stolen by young men who specialise in pickpocketing mobile phones (Little 2016). If, however, repair technicians benefit from the operation of a black market in mobile phone sales, then they also feel the negative effects on their repair business when the price of a pre-owned phone on the street is less than the cost of repairing a broken phone.

Second, I ask: How do the ethics and values of a gift economy shape the principles and practices of the informal economy that has emerged around mobile phones? Michelle MacCarthy (2011) has addressed this question with respect to the economic opportunities that mobile phone uptake has created for Trobriand Islanders. These opportunities include using generators and solar panels to recharge phone batteries at 2 kina per phone and reselling Digicel phone credit (with a small markup) sent by wage-earning relatives in town. MacCarthy emphasises how pursuit of these money-making opportunities requires entrepreneurial individuals to negotiate the logics of both market and gift exchange. On the one hand, they must 'try to escape, rather than embrace, social obligations, as relatives will try to exploit kin relationships to get something for nothing'. On the other hand, 'refusing to share freely … brings its own perils; one must carefully balance the desire to make money, and the necessity of respecting kin obligations' (MacCarthy 2011: 5–6). The Trobriand entrepreneur's dilemma consists in uncoupling transactions from kin relations just enough to generate income but not so decisively as to erode the moral basis of 'mutuality of being' (Sahlins 2011).

In other words, there is no reason to think that in PNG the presence of market transactions signals the necessary, let alone inevitable, displacement of the morality of gift exchange. Busse and Sharp's (2019: 139) ethnographies of what goes on in PNG marketplaces demonstrates 'the nuances and complications of gift and commodity exchange'. That is, what appear to be

---

8   No technicians claimed to know how to unlock phones that are tied to one network so that any SIM card could be used in the phone.

impersonal commodity transactions—the sale of betel nut or bananas for cash—often include elements that acknowledge and reproduce a personal relationship between buyers and sellers (see Chapter 2). A little 'extra', for example, might be added to the purchase by the seller. Marketplace conventions in PNG assert a morality of exchange different from the instrumental rationality associated with high-frequency algorithmic trading or the dubious ethics of the used car lot. PNG marketplaces, by contrast, are characterised by 'the relative absence of price competition, reluctance to drop prices, the absence of hard sell tactics' and the absence of bargaining and haggling (Busse and Sharp 2019: 138).

Conventions of care shape the relations of mobile phone street vendors with their customers as well as with each other. Matthew, a single man in his early thirties, repairs mobile phones and sells airtime at a street stand in Goroka. He occupies a particular spot on a particular corner with other vendors who sell cigarettes, betel nuts, hard-boiled eggs, soft drinks and the flex cards that people buy to top up their mobile phone airtime balances. Matthew was one of the first to teach himself how to repair phones; he and his friend and workmate Gabriel were also among the first to start selling airtime in 2009 as Digicel rolled out its network of cell towers across the country.

Matthew regards his work as a service to the community; indeed, to the nation. His business yields to the demands of a moral economy that EP Thompson (1971) would find familiar. Villagers who come to town with no money can offer their homegrown bananas or sweet potatoes in exchange for repair services. Town workers, however, will be sized up and charged according to Matthew's estimate of their ability to pay. Matthew likes to say that many of his customers offer him small gifts of food and soft drinks. Other vendors talked about memorising the phone numbers of regular customers, waiving service fees for direct top-up, and extending credit (*dinau* in Tok Pisin) when customers are short of cash.

Goroka street vendors, moreover, care for each other: they all offer their goods at the same price and eschew overt competition. Vendors set up their stalls side by side and offer the same flex cards at the same price. One vendor's business collapsed when he overspent his revenue and was unable to purchase a new supply of flex cards. A fellow vendor hired him until he was able, a year later, to save enough money to re-establish his own business. As Bai Magea (2019: 39) puts it, 'Airtime vendors and mobile repairmen in Goroka recognize themselves as a community'.

Bai Magea notes, however, that vendors and repairmen wonder if Digicel recognises them as part of the Digicel corporate community. Technically, the street vendors are not Digicel employees; their connection to the company is more at arms-length then it was in the heyday of the *Strit Venda* Programme. Matthew, nevertheless, insists that he and his fellow vendors are part of Digicel: without them the network would not work. Street vendors, in addition, suffer when the network falters. If the network is unstable at the moment when a vendor transfers top-up credit from his phone to a buyer's phone, then the vendor might have to resend the credit, absorbing the loss himself. Vendors also have to deal with the fallout from Digicel's promotions, as when customers angrily complain that they have not received the credit they purchased when in fact the credit had been automatically deducted to pay back an outstanding loan.

Matthew's recognition of 'people as infrastructure' (Simone 2004) is entirely plausible. Digicel, the brand name under which Telstra trades in PNG (Business Advantage PNG 2021), still relies on vendors in townships like Goroka to distribute airtime credit into the hinterlands; residents of outlying villagers prefer to buy flex cards in town for later use or for resale. In cities like Port Moresby, however, the advent of smartphones has enabled more and more people to top up directly online, buying airtime like Linda through linked banking accounts or the My Digicel app (see Chapter 4).

The future of the prepaid scratch-off flex card is dubious, and the economic niche of street vendors is shrinking (see Chapter 4). There is a policy argument to be made that preservation of the flex card vendors' livelihood would support the informal economy on which so many Papua New Guineans depend (see James et al. 2012). For Matthew, who regards his work as a service to the local community and larger nation, there are ethical as well as economic questions at stake: Does Digicel really care about him and his fellow vendors, who were present at the beginning when the company was first establishing itself in the country? Will Digicel take care of them?

There are, in the end, limits to the extent that street vendors and others who work in the informal economy of mobile phones can appropriate Digicel (or Telstra); that is, compel the company to acknowledge the morality of gift exchange that many Papua New Guineans take for granted. Matthew's concerns accordingly raise open and unsettling questions about the power relationships embedded in a new technology (Bar et al. 2016). Does Digicel still care for the people of PNG more than 15 years after the company brought them a revolution in communications? Would Telstra, having acquired Digicel, do better?

4

Smartphones and Data: Convergence and Content

The rapid introduction of smartphones into Papua New Guinea (PNG) focused consumers and companies alike on the primacy of data for internet access rather than the capacity to make voice calls or send text messages (SMS). These devices brought new opportunities for both consumers and companies to exercise freer agency in their relationship with each other. For example, consumers could not only exploit the affordances of smartphones in their own personal projects of self-cultivation (Miller et al. 2021), but also use social media to register public complaints about poor mobile service. Companies, for their part, could target users more precisely with online marketing and induce users to purchase credits and to seek customer care via apps rather than in person. Smartphones, however, also brought new challenges and constraints. For consumers, using data required new choices and calculations with regard to managing prepaid subscriptions. For companies, the turn to data meant addressing the decline in revenue from more lucrative voice services.

This chapter interprets these opportunities and challenges by deploying the contested notion of media convergence. I focus specifically on the economic and cultural dimensions of media convergence that bear upon the issue of content.[1] By economic convergence I refer to the evolution of telecom firms like Digicel into something else—companies that are simultaneously

---

1   For a discussion of media convergence in terms of four dimensions (technological, economic, political and cultural), see Miller (2011). For a related historical discussion of media convergence from the 1980s to the early 2010s, see Balbi (2017). For critical assessment of the idea of media convergence, see Silverstone (1995), Jenkins (2001) and Yrteberg (2011).

internet service providers (ISPs) and entertainment and news companies; in short, providers of content to be consumed through the use of data. By cultural convergence I refer to the participatory possibilities opened up by the advent of tools allowing ordinary people to 'archive, annotate, appropriate and recirculate content' (Jenkins 2001: 93). These possibilities have altered the ways in which companies and consumers can engage each other and perforce make claims on each other within the moral economy of mobile phones. In attending to these possibilities, I extend the discussion of technology appropriation cycles initiated in Chapter 3 and reprise the book's overall theme of freedom and constraint with respect to mobile communication.

The New Data Landscape

The spread of smartphones in PNG was underway by 2011, when Digicel introduced 3G/WCDMA mobile broadband services throughout the country. Steep drops in the price of new entry-level smartphones as well as a robust market for second-hand handsets, often stolen, helped to accelerate the spread, especially in urban areas where access to 3G towers was more available and reliable. In July 2014, Digicel was offering an Alcatel OneTouch Pixi (equipped with 3G touchscreen, 512 megabytes [MB] internal memory, 256 MB RAM and a microSD slot) for PGK149 (about USD48). This low-cost device was, of course, locked to the Digicel network.

The year 2015 was an important moment in the proliferation of smartphones in PNG due to the country's role as host of the Pacific Games. Telikom PNG, bmobile-Vodafone and Digicel all seized the opportunity to promote the sale and usage of smartphones to follow the results and to watch highlights of sporting events. Digicel offered its new Alcatel Pixi 2 for 139 kina. The phone featured 4 gigabytes (GB) of memory, 512 MB RAM, a dual core 1 GHz processor and Android 4.2 operating system— a significant upgrade from the original model that sold for 10 kina more just one year earlier.[2] Digicel COO Paul Stafford reported that sales of the Pixi 2 had skyrocketed as a result of the Games:

---

2   For the record, Vodafone PNG offered the Alcatel 1L in a Christmas 2022 promotion that also included a free basic phone with 20 kina of on-net credit for 199 kina. The Alcatel 1L features a 6.1" HD display with a quad core processor, 2 GB RAM and 32 GB internal storage.

It's clear that people in PNG see the value in the world of information, fast social media sharing and access to entertainment that smartphones open up, when running on Digicel's network, which is the fastest and most reliable in the country. (Post-Courier 2015b)[3]

In anticipation of the Pacific Games, Digicel announced a USD3 million upgrade of network capacity in Port Moresby, and a separate investment of USD13.4 million in 56 LTE sites, also in Port Moresby. The company added new towers and increased the capacity of existing towers in the vicinity of stadiums and other event venues. CEO Maurice McCarthy exclaimed:

The beauty of this investment on the network is that it will benefit the citizens of Port Moresby long after the games are over with faster more reliable internet speed and consistent coverage. (Post-Courier 2015c)

Like annual Independence Day celebrations, the Pacific Games offered Digicel a platform on which to represent the company as attentive to the needs of the nation, including the infrastructure necessary for development. McCarthy noted that Digicel's investment would allow Port Moresby to host 'many more large scale events which have high traffic periods' and which in the past have placed 'enormous pressure on the network's capacity' (Post-Courier 2015c).

Corporate officials of bmobile-Vodafone would probably have objected to McCarthy's boasting. Like Digicel, bmobile-Vodafone regarded the Pacific Games as an opportunity to promote its delivery of services as the 'official mobile services provider'. Pacific Games CEO Peter Stewart celebrated how bmobile-Vodafone handled the heavy traffic during the opening and closing ceremonies when it seemed 'like the entire nation [was] using their network at the Sir John Guise Stadium' (National 2015a). The company, moreover, opened a retail store at the international airport just in time to greet visitors for the Games, including about 4,000 athletes and officials who were issued free SIM cards. Telikom, for its part, advertised its status as 'exclusive telecommunications provider' for the Games. It also offered free wi-fi service at several of the venues for the Games, and extended the offer for a month after the Games ended. At the same time, the company promoted EV-DO and MiFi devices[4] at competitive prices and with a free 2,015 MB of data. Plainly, the move from voice and SMS messaging to

---

3   The games also provided Digicel with an opportunity to generate income from SMS subscriptions that provided users with medal results and inspirational quotes for 10 toea per text.
4   Devices for wi-fi data transmission: EV-DO stands for the name of the wireless transmission standard 'Evolution-Data Optimised', while MiFi is a small portable wireless router.

data had begun in earnest for both the residents of Port Moresby and the companies that provided them with telecommunications services. In 2016, a Digicel official estimated that the company served 700,000 to 800,000 smartphone subscribers (personal communication, May 2016).

By 2021, the rising use of smartphones was indicated by the statistics found on the website of We Are Social, a global agency specialising in social media marketing. The agency estimated some 3.11 million 'mobile connections' (not unique subscribers) in a country of about 9 million people (34.4 per cent of the population). Of these connections, 55.2 per cent were broadband connections. Of an estimated 930,000 active social media users (10.3 per cent of the population), 97.5 per cent accessed platforms such as Facebook—by far the most popular social media site in PNG—via mobile phones. But recall that the spread of smartphones indicated here is highly uneven (see Chapter 1): 'Internet usage is skewed towards urban centres, with almost 70 per cent of internet users residing in the cities of Port Moresby and Lae' (Highet et al. 2019: 24). And these users, moreover, are more than likely to be men under the age of 34.

Prepaid Subscriptions: New Challenges to Consumers and Companies

The coming of smartphones and the concomitant demand for data entrained different challenges for ordinary consumers, mobile network operators and informal businesses. Let's begin with individual consumers on prepaid subscriptions (leaving aside the minority of consumers on postpaid plans and business customers), for whom the challenge of affordability remained paramount. A 2019 survey of 638 internet users across PNG indicates that 54 per cent spent less than 10 kina per week on data while only 16 per cent spent more than 20 kina (ABC International Development 2019). Access to the internet rarely comes in the form of free public wi-fi service, which is almost non-existent in PNG; hence the appeal of Telikom's unusual offer to supply free wi-fi at selected sites during the Pacific Games. Some office workers can take advantage of internet access while on the job, using it to download and share music, for instance (Crowdy and Horst 2022), but most individuals must purchase data from Digicel, Telikom, bmobile and, since mid-2022, Vodafone PNG. Even students at the University of Papua New Guinea often prefer (or feel compelled) to purchase data rather than deal with the university's free but slow and unreliable wi-fi service.

In PNG in 2015, one could buy data credits from Digicel that could be used for one hour (10 MB for 99 toea) or for 30 days (1,500 MB for 60 kina) with several options in between (see Chapter 3). Each option, however, came with the caveat of 'use it or lose it'. Buying data bundles restricted in this way was nevertheless preferable to paying 49 toea per megabyte (30 toea in 2021), the very expensive 'out of bundle rate' charged by Digicel (see Chapter 3). Navigating the data landscape, then, meant first of all choosing which data bundle to purchase and keeping track of one's usage. Prepaid data subscribers thus confronted the problem that Donner refers to as 'the lack of transparency about units and experiences' (2015: 127):

> With voice calls and SMS messages, on the one hand, users can pay by the minute or message, and have a sense of how their behavior maps to expenses. With data, on the other hand, prepay users purchase some aggregation of mysterious bytes.

The CEO of bmobile-Vodafone, Sundar Ramamurthy, who arrived in late 2013 with the aim of making the state-owned company more competitive, sought to simplify the growing complexity of operating a mobile phone by introducing bundles that assembled data, voice and SMS in a single package. For example, in 2021 bmobile offered several different MOA packs (*moa* means 'more' in Tok Pisin), such as the 'MOA week' which for seven days included unlimited voice calls to other bmobile numbers ('On Net'), 30 minutes of calls to other numbers ('Off Net'), 30 SMS and 300 MB of data at a price of 9 kina. (Digicel began offering similar '1TOK bundles' in 2016 that included data, SMS and 'unlimited Digi to Digi talk time'). Ramamurthy explained that he preferred a single straightforward value proposition to promotions that came with conditions and in effect dictated behaviour: 'The fine print is that there is no fine print' (personal communication, March 2015).

Digicel, by contrast, continued to use promotions as a marketing strategy for socialising consumers to increased data usage. In 2014, special days were announced on which the purchase of a 'week pass' for 10 kina would double the amount of data from 300 MB to 600 MB. In April 2015, signalling how it imagined the source of future revenues, Digicel introduced a new data plan: for 4 kina, one could now buy a 24-hour data bundle (the 'social pass') that supplied not only 60 MB of data, but also unlimited access to Twitter, Instagram and Facebook. In July 2016, Digicel introduced free Facebook basics. A surprisingly candid report from Digicel's own news service (Loop PNG 2016) explained the move:

> Digicel PNG is on the path now to increase data penetration throughout the country through the use of its recently launched service, free Facebook.

> Acknowledging Facebook as the largest social media platform globally with 1.6 billion customers, Digicel PNG partnered with Facebook to provide free browsing to its PNG customers.

> Digicel managing director mobile, Shivan Bhargava, said Facebook may possibly be a starting point to get more Papua New Guineans onto social media.

Digicel announced, however, in October 2016 that Facebook Free Mode would henceforth only be available to users who top up their airtime balance by 3 kina—and then it would only be valid for two days. (For an account of the rise and fall of Facebook 'free basics' see Hempel 2018.)

Navigating the new data landscape posed other challenges besides the affordability of data. For example, data bundles often renewed automatically unless a user had explicitly chosen to opt out of this function. Users also needed to keep track of their data usage lest they be charged at the steep out-of-bundle rate. This requirement involved developing an intuitive sense of how much data would be consumed by watching a video or making a call via WhatsApp. YouTube videos were thus likened to electric jugs— they consumed large amounts of data just as kettles used for boiling water consumed large amounts of electricity. Many users experienced the disappearance of data for apparently no reason, which usually led to accusations against Digicel and a pervasive mistrust in the company's accounting procedures. In 2019, Digicel introduced rollover for some data plans. However, 'if a customer does not make a purchase within 24 hours of the expiry of a data plan, unused data is lost' (Watson et al. 2021).

While many users complained that their data was disappearing or even being stolen by Digicel (see Foster 2020), the company suggested that users must learn how to manage the settings on their smartphones, turning off the auto-sync function, for example, or remembering that music streaming might use up to 1 MB per minute. That is, Digicel responded to the concerns of its users over how the company measures usage and charges for data with exhortations to assume greater individual responsibility for operating personal devices. Digicel went so far as to offer 'Smartclinics' at its retail outlets in order to educate the public in matters of data usage and social media (National 2015b). I spoke with one consumer who responded to the demands of navigating the data landscape by inventing new forms of self-discipline (see Chapter 3). This university student claimed that she would rather buy three separate 60 MB day passes at 3 kina each instead of a discounted 170 MB 3-day pass at 8 kina in order to give herself more control over her data usage:

> When I subscribe for more than 60 megabytes I find that my data finishes very quickly, leaving me feeling bad for spending money on something that did not last for the intended number of days.

Even if a user could afford to purchase data and engage with social media, there were new communicative circumstances to consider. For example, a user's online presence might be detected by other users, thereby compromising the desired privacy associated with individualised use of mobile phones. An active Facebook user's presence online is indicated by a green dot (*'Green dot stap'* in Tok Pisin) next to their profile image. One can send a credit request to the active Facebook user; if there is no reply, then the requestor knows that they are being ignored. Similarly, although calls made by WhatsApp are cheaper than conventional voice calls, they require the recipient as well as the caller to use data. Whether one chooses to respond to a WhatsApp call now involves the same kind of cost–benefit calculation involved in responding to a 'call me' or 'credit me' request (Chapter 3).

It is important to note that one popular strategy for navigating the data landscape amounts to avoiding it. That is, one can enjoy the multiple affordances of a smartphone without using, let alone purchasing, data. Consider the case of Marie, a grandmother in her fifties who works as a housekeeper in Port Moresby. Marie took advantage of a 2-for-1 sale at a Digicel retail shop to purchase Alcatel OneTouch phones for herself and her daughter. Marie's daughter assists in topping up the phone from Marie's bank account and then transferring a few kina from Marie's phone to her own. Marie herself uses her phone to make and receive calls, and to receive but not send text messages. She does not use the phone to go online. Marie does, however, use several other of the phone's features. She takes and stores photos of her family; she listens to gospel music; she reads the Bible that her daughter put on the phone; and she records sermons at her Assemblies of God Fellowship Sunday services and listens to them afterwards. The phone is an important asset to Marie's spiritual life as well as a means for staying in touch with her home village located about 70 km away.

Geoffrey Hobbis (2020) has documented extensively the ways in which rural Lau Lagoon villagers in Malaita, Solomon Islands, have embedded smartphones in their everyday lives despite lacking the financial means to make frequent calls and the infrastructural capacity to access the internet (cf. Tenhunen 2018: 157). Lau villagers make full use of the multimedia and computational affordances of smartphones, watching videos, listening to music, playing games, doing calculations and sharing photos (as well as using the flashlight). These offline uses of the phone constitute in one sense an appropriation of the network. Like Marie, Lau villagers take advantage of the subsidies that telcos give to handsets in the expectation that the spread of inexpensive handsets will stimulate data usage. The frustration of this expectation causes concern among telco officers. Then Digicel CEO John Mangos, questioning the logic of subsidising phones when people were not going to use data, acknowledged in 2015 that consumers like Marie were 'a problem' for the company (personal communication, 19 March 2015).

From the company's point of view, however, the transition to data was more than a question of whether to subsidise handsets. In 2017, researcher Amanda Watson asked Gary Seddon, a senior manager at Digicel, about the threat that voice over internet protocol (VOIP) posed to the company's income from voice calls and text messages. Seddon replied:

> It encourages data consumption, so you will see greater data growth. It is a growing medium of communication in PNG; one that we embrace. But let's not forget that conventional voice and SMS services are still incredibly important in PNG. The populations that are outside of 3G and 4G zones still rely heavily on our 2G based services. There remains a lot of the country to appropriately cover, despite Digicel's considerable investment, presently exceeding PGK2.5 billion. (Watson and Seddon 2017)

Seddon's comments point to an emerging digital divide in PNG: data consumption is increasingly concentrated in cities like Port Moresby and Lae, whose residents expect faster and cheaper broadband connectivity for their smartphones. Seddon speculated that meeting future internet bandwidth demands would inevitably require incorporating 'off-island fibre optic delivery methods' (Watson and Seddon 2017). In other words, the transition to data posed an infrastructural challenge for a company that had hitherto obviated the necessity of dealing with the monopoly on submarine gateways held by state-owned companies (first Telikom, then PNG DataCo) by securing access to bandwidth via O3B satellites (see Chapter 1). As Seddon predicted, reductions in the wholesale price of bandwidth due to the launch of the Coral Sea Cable System were announced in early 2021 (Watson et al. 2021).

From the point of view of mobile network operators, the transition to data also posed the challenge of free riders on the network—not wily users devising clever workarounds, but huge multinational corporations such as Facebook and Google. These companies not only provide apps that allow users to make calls and send text messages, but also derive revenue from the traffic generated by their ads. In 2015, Denis O'Brien vented his frustrations over such appropriation in several media outlets. The *Financial Times* (Cookson 2015) quoted O'Brien as saying:

> Companies like Google, Yahoo and Facebook talk a great game and take a lot of credit when it comes to pushing the idea of broadband for all—but they put no money in … Instead they unashamedly trade off the efforts and investments of network operators like Digicel to make money for themselves.

O'Brien famously told the *Wall Street Journal* that 'Mark Zuckerberg is like the guy who comes to your party and drinks your champagne, and kisses your girls, and doesn't bring anything' (Knutson and Schechner 2015). He claimed that Google earns 'billions of dollars on advertising, and they don't pay a penny. I think it's the most extraordinary business model in modern history' (Knutson and Schechner 2015). It is a business model that offers no incentive for mobile network operators to invest in infrastructure, especially in places like PNG where such costs are extremely high.

O'Brien went so far as to begin blocking ads in Jamaica, partnering with the Israeli company Shine Technologies whose software 'prevents online ad networks such as those operated by Google from delivering display and video ads to mobile browsers and apps' (Cookson 2015). O'Brien justified his decision as serving the best interests of consumers. Digicel asserted that ads use up to a tenth of a user's data plan allowance, and that eliminating unwanted ads would save money for consumers. The ad-blocking initiative met some regulatory resistance in a few Caribbean markets (Paul 2015) and in Europe (Irish Times 2016), while both Google and Facebook attempted to address the concerns of mobile network operators by supporting various efforts to get more users to use more data. These efforts included both offers of free access to a simplified version of Facebook (Facebook Zero or 'free basics') and investments by Google in submarine fibre optic cables as well as O3B satellites and hot air balloons designed to bring internet access to remote locations. In 2017, Shine rebranded and ceased selling network-level ad-blocking software.

The transition to data, finally, posed challenges to the human infrastructure of telecommunications networks in PNG. That is, the spread of smartphones and increase in internet use threatened the livelihoods of street vendors (Chapter 3). In cities and towns, the advent of smartphones enabled more and more people to top up directly online, buying airtime through linked banking accounts. Moreover, new forms of 'self-care'—quite different from the kind of care that street vendors give their customers (see Chapter 3)— are being promoted. In January 2017, Digicel launched the My Digicel app for smartphone users, promising customers an efficient tool for 'managing their Digicel life' (National 2017b). Several months later, the company introduced a menu that allowed customers, including users of basic handsets, to assist themselves with queries relating to data and top-up, among other things. Dial *123#. 24/7. Free. A list of frequently asked questions appears on the phone's screen.

Media Convergence (Economic)

Digicel embraced digital media convergence, evolving into a provider of content as well as connectivity in all its markets. Across the Pacific region, the company diversified into the consumer entertainment business, launching its own television network and acquiring cable and satellite television broadcasters and ISPs. In PNG, state-owned Telikom did likewise. Its purchase in early 2015 of EMTV, the country's major free-to-air commercial television service, was a clear signal that the mediascape in PNG was being radically reconfigured.

Digicel's plan to become something more than a mobile network operator was indicated in the US Securities and Exchange Commission (SEC) filing for the company's aborted initial public offering in 2015. The Form F-1 Registration Statement (Digicel Group Ltd 2015) outlines a vision of the 'Digicel Ecosystem':

> The Digicel Ecosystem promotes customer retention and aims to drive revenue growth by capturing customers' spending on communications and entertainment across all of Digicel's businesses. It is a platform to which new products and services can be added. Importantly, Digicel has developed a strategy around proprietary content, which includes SportsMax, with exclusive sports content options tailored to individual markets, Loop, a local news and content app that is currently the most downloaded news app in the Caribbean with 450,000 mobile downloads as of June 9, 2015, and

mobile financial services through Digicel Mobile Money, Boom and Bima, which provide mobile banking and micro insurance services in various markets.

This ecosystem presupposes, of course, that Digicel is an internet company. In PNG, the company's move in that direction first included the acquisition in 2011 of DataNets, an IT services firm founded, ironically, by Sundar Ramamurthy. According to a report from Oxford Business Group (2014), in 2013:

> Digicel purchased the very-small-aperture terminal business of Remington Group, a locally-owned diversified conglomerate, and then later in the year took over as the wholesale provider for Daltron, a local ISP that is part of the WR Carpenter Group. In sum, the company, which already dominates the mobile phone business, has significantly expanded its presence as an internet company.

Telikom PNG, for its part, acquired Datec, a wholly owned subsidiary of Steamships Trading Company, in 2014:

> Datec provides a wide range of ICT services, solutions and products in networks, applications, training and web design. Datec will be used in part to help Telikom PNG expand into internet and data services. (Oxford Business Group 2014)

At the same time that Digicel expanded into internet and data services, it also began to reshape itself as an entertainment company. In 2014, Digicel acquired the cable TV operator Channel 8 in PNG as well as Hitron Limited, in which Digicel took a 60 per cent shareholding interest. According to the 2015 SEC filing (Digicel Group Ltd 2015):

> Hitron provides Cable TV and internet service provider ('ISP'), services across a multichannel multipoint distribution service network in Port Moresby, wholesale content services to some of the small cable operators and remote communities such as mining camp [sic]. It also implements and maintains VSAT systems for remote areas.

In the same year, moreover, Digicel launched Digicel Play, a television service offering 29 channels including not only CNN and MTV but also TVWAN, a locally produced free-to-view channel. Customers could access Digicel Play by buying a Digicel Play Box for 169 kina and connecting it to their television set. Digicel Play offered three free-to-view channels. Access to the pay-to-view channels required prepaid pay-as-you-go subscriptions just like the prepaid subscriptions for mobile phones:

> Anyone connected to the Digicel Play platform can sign up at any
> time for a Pre paid package to suit their budget and viewing needs.
> The Premium Package offers the best value for money at just K99 for
> 30 days, with full access to all 29 channels. Other options include a
> K20 for 7 days plan with access to 8 channels and a K40 for 30 days
> plan with access to 16 channels. (Loop PNG 2015)

The same flex cards used for prepaid mobile phone subscriptions could be
used to pay for Digicel Play. As CEO John Mangos explained in 2015,
Digicel was applying the same model for phone credit sales to TV (personal
communication, 2015). Mangos said that he expected no more than
100,000 households to acquire a Digicel Play Box. Digicel would at first
transmit using its towers, rolling out the TV service in Port Moresby and
Lae and large towns and eventually obtaining satellite space and making
the service available to the whole country. In other words, Digicel had
become a telco in the prepaid subscription TV business, a rare if not wholly
unique arrangement.

Telikom PNG, which already owned Kalang Advertising Limited, operator
of FM100 and HotFM 97.1 commercial radio stations, matched Digicel's
move into television by acquiring EMTV, the country's most viewed free-
to-air television station. Although there were reports at the time that Digicel
was seeking to acquire EMTV, Telikom eventually purchased the station in
2015 for USD10 million from Fiji TV, owned by Fijian Holdings Limited.
(Digicel, however, acquired for TVWAN the coveted rights to National
Rugby League [NRL] broadcasts that EMTV had previously enjoyed; see
Chapter 2.) *The Australian* reported that Telikom's CEO, Michael Donnelly,
attributed the value of the acquisition to convergence, 'the opportunity to
channel increased content and other services to its customers' (Callick 2015).
Bhanu Sud, CEO of EMTV, echoed Donnelly's observation: 'The PNG
media market is shifting thanks to the convergence of different technologies,
and multimedia companies are competing not just domestically, but across
the Pacific region as a whole' (Oxford Business Group 2015).

By 2017, when Amanda Watson interviewed Mahesh Patel, Telikom's board
chairman, the fact and value of media convergence had become taken for
granted. Patel told Watson (Watson and Patel 2017):

> In international trends, it's all about combined services. Our
> competitor already has a television service. So Datec and EMTV
> add value. Datec is a wholly-owned subsidiary as well. It's an internet
> service provider, providing services into homes. EMTV is screening

NRL games again. Our competitor has the rights, so we're running delayed telecasts. We are one of the top-ranked local content providers in the world—60 per cent of our EMTV content is local.

Both Telikom and Digicel had undeniably become 'content providers' in addition to telecommunications and internet companies.

Prominent among the sundry content that Digicel and Telikom PNG sought to provide were news and information. The barriers between print news and digital news began to fall quickly once social media users began posting photos of newspaper articles to Facebook and conversely the daily print newspapers reported viral social media posts. EMTV news, including video of television broadcasts, remains available on the company's website, Facebook page, YouTube channel and Twitter and Instagram feeds. Digicel's counterpart to EMTV news was its free-to-browse news service, Loop Pacific, launched in 2014; an app was launched in November 2015 in PNG. *The National* then reported that 'Anyone with a Digicel SIM would be able to visit the Loop Pacific app or website from their Apple or Android smartphone without using any credit on their account' (National 2015c). Digicel Pacific customised Loop Pacific news for its markets in PNG, Nauru, Samoa, Tonga and Vanuatu (but not Fiji).

A 2019 report on media engagement suggests how smartphones have changed the media landscape in PNG. The vast majority of PNG citizens surveyed (n = 638) who access the internet do so through a mobile phone, and a large proportion use the internet for news: '*Loop PNG* is the most popular source of news, with weekly reach of 29% among regular Internet users' compared with 14 per cent for EMTV Online (ABC International Development 2019: 40). Loop's popularity perhaps stems from its free availability to all Digicel subscribers. Not surprisingly, television viewing of EMTV has declined as viewers migrate to online news sources where internet connectivity is accessible. Focus group participants claimed that 'DVDs are no longer used, having been replaced with phones, memory cards and flash drives as well as pay TV (Digicel Play) and movie programs' (ABC International Development 2019: 22).

The expansion of Digicel's ecosystem and its corporate evolution since beginning as a mobile network operator culminated in an October 2020 proclamation declaring the relaunch of Digicel as a 'Digital Operator'. Digicel was going 'all in on digital' (Loop PNG 2020):

Oliver Coughlan, Digicel CEO Caribbean and Central America, explains: 'Today, we're stepping into our future. As a mobile operator, we sold minutes and MBs, now, as a Digital Operator, we're about delivering digital experiences. And we're making that happen 1440 minutes of each day. That's all day, every day'. (Loop PNG 2020)

These digital experiences were accessible in the form of prepaid 'Prime Bundles':

Digicel Prime Bundles feature all of Digicel's suite of digital services (or apps) spanning **D'Music** for music, **PlayGo** for TV streaming, **SportsMax** for all things sport, **BiP** for advanced messaging, video and voice calling, gaming and marketplaces, **LOOP** for local and international news, **GoLoud** for 75 local radio stations and podcasts and **Billo** for cloud storage—each with its own super generous data allotment so that customers can feel, touch, experience and enjoy them as they live their digital lives. And on top of that there's Digicel's self-care destination, **MyDigicel app**.

Not only that. On top of those digital services, each of the Digicel Prime Bundles is loaded with a healthy helping of any use data for customers to enjoy. (Loop PNG 2020, original emphasis)

Prime Bundles signalled not only a decisive break with past practices of selling minutes and megabytes, but also a new definition of the company–consumer relationship, one in which the power relationship embedded in the smartphone heavily favoured Digicel (see Chapter 3). Instead of appropriating a Digicel device as the means to cultivate one's social network and personal interests, Digicel would effectively incorporate the user into the company's proprietary content, thus appropriating the user by means of the device.

***

It is worth making a brief note here about political or regulatory convergence. In PNG, there are no constraints on media cross-ownership or foreign ownership of media. The establishment of the National Information and Communications Technology Authority (NICTA) ended the dual regulatory arrangement that created the tussle between the Papua New Guinea Radio Communications and Telecommunications Technical Authority (PANGTEL) and the Independent Consumer and Competition Commission (ICCC) over Digicel's entry into PNG in 2007 (see Chapter 1). As a converged regulator, NICTA 'can license operators, undertake inquiries, determine interconnection prices, make recommendations to the Minister to declare (regulate) specific services, and oversee a Universal Access Service (UAS) program' (Howell et al. 2019). NICTA does not issue service-specific or technology-specific licences. For example, Telikom was able to use for its 3G network the spectrum originally allocated for Citifon's CDMA network (Watson et al. 2017). Instead, NICTA supplies licensed operators with application, network and content licences (in addition to radio communications and cabling licences). In July 2011, 'NICTA granted Digicel three new ICT [information and communications technology] licenses—Network, International Gateway, and Application Licenses— under the new National Information and Communications Technology Act, 2009 (National ICT Act)' (Digicel PNG 2012). Digicel was the first mobile operator to migrate to the new ICT licensing regime. It was also 'the first mobile operator in PNG to successfully migrate its spectrum licenses to the National ICT Act by signing a new Spectrum Usage Agreement (SUA)' (Digicel PNG 2012). In April 2012, Digicel received from NICTA the content licence required for exploring internet protocol television options.

The fact that NICTA's funding comes entirely from licence fees rather than government budget implies a potential conflict of interest, given that operator fees are proportional to firm revenues (Howell et al. 2019). NICTA also depends on Digicel for the data needed to make decisions about matters such as taxes or termination rates or the use of the UAS fund. Digicel's continued expansion across media and its sheer dominance of the market in PNG presented regulatory challenges to an authority established in order to promote competition.[5] Digicel's partnerships with members of parliament for building towers (see Chapter 1) also made helpful allies within government. According to NICTA officials with whom I spoke, the company's sense that 'we created the market' informed Digicel's reluctance to share it, for example, by sharing its infrastructure with competitors. Inside NICTA, Digicel was sometimes referred to as 'the red brigade'.

At the same time, technological convergence delivered its own challenges. In a 2017 interview with several NICTA officials, Amanda Watson (Watson et al. 2017) asked: 'Turning now to convergence of different technologies into single devices, what kind of issues does this present to NICTA?' To which the NICTA officials replied:

5   In this regard see Watson and Fox (2019), who claim that 'There appears to be no publicly available information on how much, if any, company tax it [Digicel PNG] pays' and accordingly urge the company to voluntarily provide this information.

> One of the big issues we've been talking about is 'over-the-top' applications—anything that uses the infrastructure and Internet protocol to provide a service. That includes broadcasting over the Internet, competing with TV and radio stations, and also 'Voice-over-Internet-Protocol' or 'VOIP' in short, such as Skype. So we are mindful of that and may have to come up with some kind of regulation to ensure we protect our operators. We don't want to be stopping progress but we do have to have something in place so everyone is happy. It's a hot topic around the globe too, regulating 'over-the-top' services—it's quite a challenging issue.

This tentative reply suggests how the disruptive effects of digitisation have not yet sorted themselves out in PNG for either companies like Digicel (and its new owner, Telstra) or the state agencies nominally responsible for monitoring these companies.

Media Convergence (Cultural)

If economic convergence is a top-down, industry-driven process, then cultural convergence is a bottom-up, user-driven process. Cultural convergence refers to the participatory possibilities opened up by digital technologies that allow consumers to be producers as well. While media companies have been busy acquiring content to deliver to consumers, consumers have been busy not only commenting on, modifying and subverting that content, but also making original content of their own—from the earliest posts of bloggers to the latest videos on TikTok. This capacity for 'produsage' (Bruns 2008) has altered both the ways in which companies and consumers interact and the role of state agencies in regulating online activity conducted through mobile phones. Produsage has also ushered in a new and paradoxical phase in the technology cycle of appropriation discussed in Chapter 3, one in which users effectively pay rent in order to access content—including social relations—that they themselves have generated and cultivated.

Produsage in PNG takes a variety of forms, many of which Digicel itself instigated by eliciting user engagement. Such elicitations were part of the company's modus operandi since it arrived in PNG, sending out SMS reminders, promotions and advertisements frequently enough to annoy many users (Chapter 2). But the era of smartphones gave users the capacity to respond to Digicel in their own words. To take one simple example, Digicel's online Loop PNG news service enabled readers to comment on articles.

Digicel PNG's Facebook page allowed users to communicate with the company, and even elicit a response to their response. Some comments on the Digicel PNG Facebook page were responded to individually. This level of attention was certainly evident in 2015, when the page was relatively new and smartphone use was just gathering momentum. Digicel posts about interruptions or changes in service would regularly elicit complaints about unfair or inexplicable charges to which Digicel would respond with an invitation to forward more details to an email address.[6] In 2021, comments would also sometimes receive a personalised response, but at other times the response appeared to be generated automatically. For example, queries about difficulties logging into PlayGo would receive the following response:

> If you are unable to access Playgo App, try this to improve app experience and retry.
>
> Go to phone settings, then to App, scroll and search for Playgo. Then click storage. Once you're in, click to clear cache and data. Then you can go back to home screen and try login again.
>
> Regards,
> Digicel PNG

One commenter replied to this response by asking 'You think we don't know that?' Such surly comments rarely received acknowledgement from Digicel.

Smartphones and the transition to data advanced Digicel's goal of decreasing calls made to the support centre. Instead, Digicel engaged users through apps, and then used the data generated by this engagement 'to set up user journeys and effectively target our customers' (SWRVE 2020). Digicel enlisted the help of SWRVE, a company that describes itself on its website as:

> the world's leading CX [customer experience] engagement vendor. We have a simple goal: helping our customers know every user, anticipate their needs, and interact in the right moment, with the right message, in the right channel. (SWRVE n.d.)

SWRVE is also the name of the platform that allowed Digicel to send targeted push and in-app messages about promotions and prepaid plans in a way that served 'to reinforce previous desirable user behavior (like a plan purchased, etc.)' (Lola Akinyinka, Director of Digital Products, quoted

---

6   pngcare@digicelgroup.com.

in SWRVE 2020). Like the endless series of contests and giveaways that Digicel launched in PNG over the years (Chapter 2), SWRVE keeps users busy taking care of themselves and, more importantly, tethered to Digicel. The deployment of SWRVE by Digicel resulted in a 52.1 per cent increase 'in Day 30 user retention, driven by messaging sent to the right users at the right time' (SWRVE 2020).

Automated marketing of the kind exemplified by SWRVE—app-propriation, so to speak—might be the latest frontier in the cycle through which consumers and companies struggle over the terms of technology use. Consumers, however, now have other ways to use their smartphones and data plans to engage if not companies, then at least each other around shared matters of concern. Users can create their own Facebook pages to address issues that they feel are not being addressed by the company in question. In PNG two such pages have provided a forum for users airing grievances, issuing denouncements and seeking redress from alleged abuse by Digicel: Digicel Complaints Group and Digicel PNG Quries (sic) and C/Care. The latter page describes itself as follows:

> Here is a group created for you (Public) to share your thoughts on how you feel against our major communication company Digicel Png. Just SHARE or ASK on this group and see if we could attract the Digicel worker or any Digicel rep, who can bring it to the digicel attention.

Indeed, the page is full of queries and complaints, many of which appear to be responded to with the following advice from Digicel:

> Please post on the correct Digicel page, link shown below, you will be better assisted from there, This is not the correct page, hence you will not be assisted by anyone.
>
> www.facebook.com/digicelpng
>
> or as an alternative please use the Digicel Live Chat service for fast and efficient help.
>
> You can access it through the My Digicel app (MDA) or the Digicel PNG website. Please try it now.
>
> Let me be very clear, this is an unofficial page and anyone seeking help here will only get negative comments.

Digicel has an official page that has a dedicated team ready to assist with social media complaints. Please go through the right channels to get help.

Thank you
Digicel
Telecommunications Company
Port Moresby, Papua New Guinea

Digicel PNG Quries and C/Care claimed 6,500 members in 2021. By contrast, Digicel Complaints Group (DCG) claimed as many as 43,000 members during the peak of the page's popularity in 2016 (by 2022, however, the page was no longer active). These members, judging by their posts in both English and Tok Pisin, were largely school-educated urban men and women with the financial resources required for regular use of a smartphone. They constituted an important public in PNG, but certainly not 'the public' in a country where the vast majority of people are rural and poor, and the literacy rate for people aged 15 years and above was about 62 per cent in 2010 (United Nations Development Programme 2019). Nevertheless, the activities of this public offer insight into the dynamics of convergence and the moral economy of mobile phones that in PNG brings consumers, companies and state agencies into constant contact with each other.

DCG was established in order to record complaints that might be brought to the attention of not only Digicel but also various regulatory agencies of the PNG state, such as NICTA and the ICCC. Its mission statement exhorts followers:

Air all your complaints in this group so we can keep a track of complaints against Digicel (PNG). Also to get their attention and that of responsible regulatory bodies such as ICCC, NICTA and the PNG Government to take action on unethical and unfair business practices.

The group has also functioned as a place to share tips and advice on mobile phone use, including news about promotions and ideas about how to work around the constraints of managing prepaid subscriptions (see Chapter 3).

DCG members were particularly concerned about the conversion of prepaid airtime into data and the allegedly predatory ways in which Digicel adds and subtracts data from an individual's balance. Complaints often described the failure of credits to appear on a person's phone after a purchase has been made or the disappearance of data from a person's balance even when the phone is not in use. This 27 February 2016 complaint to the group was fairly typical:

> At exactly 11am today, I entered two K5 flex numbers: 01 7249 490 5910 & 16 2662 659 3637. At exactly 11:14am, Digicel sent me two messages – 1. Advised that I have used up my data and 2. Asked whether I need airtime of K13 advance. I immediately checked my balance only to see K5.03. I texted Digicel and 5 minutes I was reimbused [sic] K3 and not the whole K4.77. What a day light robbery!

A 25 November 2019 post similarly complains:

> Digicel, please give my credit back. This is not fair. I already paid for social pass n yet you took away all my remaining credits (K9.00) in my phone. Give it back 😢. Sick off [sic] this. This is not tge [sic] the first time.

This post elicited 22 comments, ranging from 'I am experiencing the same. That is stealing' to exhortations to switch to bmobile or Telikom.

Group members sometimes posted screenshots of the before-and-after balances on their phones as evidence of stolen data. Martyn Namorong, a prominent PNG netizen (Capey 2013), even created a YouTube video in 2015 titled 'Watch how Digicel PNG is ripping us off', in which he demonstrates the mysterious disappearance of both data and credit from his phone within minutes of topping up.[7] Such accusations of robbery and theft, of being 'ripped (off)', are frequent reminders of EP Thompson's (1971) well-known account of the protests that erupted as a moral economy of food provision gave way to practices and principles associated with 'free trade' in eighteenth-century England. These protests, which often led to direct action, appealed to notions of a just price in the face of concerns that the poor suffered at the hands of those with 'command of a prime necessity of life'. Much like the folks about whom Thompson wrote, DCG members express intense feelings over 'weights and measures' and petition the authorities—NICTA and the ICCC—to regulate business transactions.

More generally, complaints signalled a moral failure on the part of the company to meet its obligation to treat its customers fairly. For example, many group members made it clear that they thought Digicel's

---

[7] Available at: www.youtube.com/watch?v=W5e3Hds4lIM, accessed 15 August 2022.

announcement in September 2016 that the company was now charging voice calls at 1 toea per second at all times to be deceptive. While the new rate lowered the previous peak rate of 79 toea per minute, it actually raised the off-peak rate from 49 to 60 toea per minute. In 2018, one complaint to the group focused on the wide disparity between data prices in PNG and Fiji. A commenter on this post speculated darkly that Digicel is 'able to go cheap in other markets because their PNG business makes enough money to subsidize and support their operations in the rest of the Pac[ific] islands'.

A high-ranking marketing executive confirmed that Digicel Pacific was aware of the Facebook complaints and in fact worried about the mistrust that they evince (personal communication, 2016). Digicel employees (and bmobile-Vodafone employees, too) posted to the DCG page on occasion, directing complaints to Digicel's own official Facebook page. One Digicel manager shared a comment that reminded the group of what life was like before the company entered PNG:

> We complain too often about charges and services but how would your life be if Digicel had never come to PNG? What was the Governments blueprint in developing connectivity with the remotest of parts in PNG and using that as an instrument to deliver lifesaving services to remote parts of PNG??

But the company's primary response has been to encourage personal responsibility for monitoring data usage, issuing newspaper advertisements that instruct mobile users on how to save data by turning off the auto-sync function and restricting background data. The fact that there appeared to be far less activity on the DCG page in 2018 than in 2016 suggests that such consumer education, however acquired, might have helped to reduce the volume of accusations made against Digicel.

***

The opacity of data charges contributes to what Benson and Kirsch (2010) call a politics of resignation in which DCG members largely express an inability to change things—their agency constrained by the terms and conditions of their online lives (see Chapter 5). Some group members do appeal for help to NICTA and the ICCC—agents of the state that initially failed in meeting its obligations to provide an efficient and affordable telecommunications service to the people of PNG. But many other posts take up a position of ironic distance and communicate a 'general feeling of disempowerment' (Benson and Kirsch 2010: 460)—a sense of resigned

inevitability of the sort conveyed with a shrug by the American term
'whatever': 'The everyday politics of resignation implies recognition not
only that things have gone awry but also that one is practically unable
to do anything about it' (Benson and Kirsch 2010: 468). For example,
after claiming that Digicel is not monitoring the DCG page and referring
complainers to the company's own Facebook page, posters drolly liken
the group to a *haus krai*, the Tok Pisin phrase for a wake or gathering
of mourners.

On the one hand, the participatory aspect of digital media gives users some
capacity to hold companies accountable to promises of service and even
to push back against the latest developments in automated marketing.
Consider the April 2021 post to DCG in which a dissatisfied Digicel user
included a screenshot of a personalised message offering the user a free
1-hour 15 MB data plan. The user asked what 15 MB could be used for:
'Seriously, how could I stay connected with family, wantoks and the world
using the 15MB. What kinda sorcery is this?' The accusation of sorcery is
fitting inasmuch as Melanesian sorcerers and witches represent an extreme
form of antisocial selfishness, a desire to consume rather than to share
(see e.g. Munn 1986). Hence the supportive comment of another poster
dismayed by the company's lack of generosity: 'What's 15mb compared to
the millions Digicel earns in minutes?'

On the other hand, there are inherent limitations to online whining. The
questions posed above are, after all, rhetorical; Digicel was neither likely
nor expected to respond. Moreover, there is a cruel irony to this sort of
participatory culture, one that was not lost on members of the DCG. Not
only do members urge their fellow members to switch allegiance to bmobile,
they also remind each other that they are paying Digicel to complain about
Digicel. Their irony indirectly acknowledges one of the painful realities of
the new economy of digital media: that participants frequently pay rent
in order to gain access to 'content' that their own produsage has created
(see Foster 2007, 2011). Such is the nature of Facebook for many users
in PNG: 'staying connected with family, wantoks and the world'—that is,
sustaining online social relations and shared histories of one's own making—
requires first paying a toll to a mobile network provider.

Part III.
Delimiting Mobility: Regulation and Responsibility

5

Mobile Disruption: Regulation, Surveillance and Censorship

In many places throughout the Global South, the introduction of mobile phones disrupted conventions for interpersonal relations between spouses and between parents and children, while at the same time affording new means for friendship, romance and intimacy (for striking ethnographic examples from India, see Doron and Jeffrey 2013; Tenhunen 2018; from Bangladesh, see Huang 2017). Smartphones additionally afforded not only online access to illicit content such as pornography, but also new opportunities for protest and critique of government on social media platforms such as Facebook. In other words, mobile phones came to be perceived and used as instruments of disruption—both domestic or private and political or public.

Such disruption defines a productive site for describing how the moral economy of mobile phones entangles consumers, companies and state agents with each other. Papua New Guinea (PNG) state agents responded to disruption with regulatory initiatives such as cybercrime legislation that raised questions about freedom of expression and rights to privacy. Another state-led initiative, SIM card registration, obliged mobile network operators (MNOs) at their own expense to carry out statistical work that the state itself proved incapable of accomplishing. Parents similarly invented their own rules for controlling their children's use of mobile phones and social media. Each of these regulatory initiatives exposed and sometimes reproduced the deep ambivalence that many people felt about the moral dimensions of mobile communication (Lipset 2013).

In this chapter, I accordingly trace the dialectics of freedom and constraint as they played out privately in domestic relations and publicly in political practice. I thus regard mobile phones as a 'disruptive technology', but not in the business sense popularised by Bower and Christensen (1995). Rather than attend to how the mobile phone might have shaken up industries such as coffee farming or taxi services (see Chapter 3), I instead identify how mobile phones arouse suspicion about the connection between particular numbers and particular persons as well as mistrust in general about a wide range of social relations. Specifically, I describe the disruptive effects of mobile communication on marital relations and of mobile-enabled access to social media on relations between state agents and ordinary citizens. I ask: In what ways did mobile phones offer new freedoms for public expression and private pleasures, and how did these new freedoms invite new forms of regulation?

Domestic Discord

The earliest studies of mobile phone use in PNG reported that the number one concern of users was the potential for phones to promote sexual promiscuity and to break up marriages (Sullivan 2010a, 2010b; Watson 2011). A 2012 newspaper story reporting the death of one man and the injury of several others in Kiriwina (Trobriand Islands) gave credence to these concerns. The story recounted how the husband of an 'alleged adulteress' confronted and killed the man he suspected of having an affair with his wife after finding and reading 'certain text messages' on his wife's phone (Post-Courier 2012c). Most reports, however, link mobiles with gender-based violence against women. For example, a GSMA study of women with little income found that:

> Women fear receiving calls from 'unknown callers' because this might aggravate their husbands. She might be beaten or the phone smashed as the husband will fear promiscuity. Suspicion of affairs is high. For this reason some women refuse to own a mobile, as they are frightened of what will happen with their husband if they do. (GSMA 2014: 31)

Hence the finding of Ketterer Hobbis (2018) that women in rural West New Britain Province actually preferred that the nearby cell tower, torn down in a dispute over land ownership and rights to royalties from the site, remain dysfunctional. Without the tower, villagers 'did not have to frequently deal with the violence that mobile phones can bring' (Ketterer Hobbis 2018: 61; cf. Taylor 2015 for Vanuatu).

The potential of mobile phones for domestic discord is partly and unsurprisingly a function of the novel capacity for private one-to-one communication afforded by mobile phones. This capacity can readily breed mistrust, as a quick review of the ethnographic record outside PNG reveals. Doron and Jeffrey (2013), for example, report how mobile phones enabled couples in India to make initial connections and plan meetings, thereby circumventing the control of the prospective bride and groom's elders. After marriage, mobile phones enabled new brides to maintain regular connections with their natal household, thereby potentially challenging the authority of both the bride's husband and mother-in-law: 'The mobile phone was viewed as an object of distrust, unless it was monitored by the husband and family' (Doron and Jeffrey 2013: 175). Mobile communication reshaped courtship practices, conjugal relations and kinship ties, sometimes reinforcing and sometimes upsetting established conventions.

Julie Soleil Archambault's (2017) ethnography of mobile communication in Mozambique similarly suggests that phones are instruments of mistrust in the eyes of young adults (see also Kenny 2016 for Tanzania). These devices testify to deceit and unfaithfulness in the form of intercepted calls and text messages. Mobile phones—the devices themselves—thus bring out into the open secrets that ought to remain discreetly hidden. In a decidedly fetishistic sense, 'it is quite literally the phone, rather than unfaithfulness, that is understood to generate conflict and break-ups' (Archambault 2011: 452).

Similar attitudes of mistrust, laser-focused on the phone itself as the material repository of evidence of wrongdoing, have been described by Kraemer (2015) for young adults in Port Vila, Vanuatu. In PNG, young men artfully use mobile phones to manage friendships in less than transparent ways. Consider as one example, observed by a research assistant in 2016, a young man who owns three mobile phones, two of which are dual SIM phones that accommodate Digicel and bmobile cards:

> He texts me with the same number all our friends have. So one night we were sitting down telling stories and his crush (a woman) texts me and my other girlfriends asking if his phone was off because her messages were not delivered to his phone. Before replying to her text, I asked my girlfriend if she received the same text, which she

confirmed. I turned to our male friend and told him to text his crush back. He laughed and said 'I left the SIM card at home. Unlike you all, she does not have this number.'

Our research assistant expressed her dislike for how this young man was treating his crush. The incident, moreover, provoked a troubling reflection: What if most young men do the same thing?

It is important, of course, to disavow any technological determinism here. Tenhunen (2018) argues that in rural West Bengal, the availability of mobile phones has enlarged women's role in marriage negotiations. That is, the marriage system legitimates mobile phone use rather than being disrupted by it. Lipset's (2013: 341) ethnographic research also makes it clear that Sepik area villagers and peri-urban residents in PNG welcome mobile phones approvingly when put in the service of 'collectivist values', for example, when used to strengthen 'the kinship-based networks they hold dear'. It is only when mobile phones facilitate 'ego-centered networking', specifically, arranging sexual liaisons (which, of course, are hardly new occurrences), that they attract moral opprobrium.

Similarly, it is important to resist the fetishism of Archambault's interlocutors. The ethnographic record demonstrates how mobile communication, rather than provoking conflict, can provide the means for overcoming the mistrust that threatens certain intimate relationships. Juliet Gilbert (2016: 3) discusses how young women in Calabar, Nigeria, manage fears of female friendships that they experienced and imagined as 'tainted with acrimony, jealousy and backstabbing behavior'. Such wariness of social intimacy can be reduced by mobile communications that allow young women to keep in touch with their 'contacts', but also to maintain a safe distance from each other. Brief calls and short text messages, including Blackberry Messenger Service messages, engendered a satisfying kind of intimacy among contacts, not friends, who were practically strangers:

> It is these forms of affect and entertainment, and the social rule of never giving exact information about oneself, which are central to understanding how young women not only alleviate boredom in the house but also manage contacts discreetly. Furthermore, just as these styles of mobile communication generate affect free from fear between young women, they also enable young women to chat with men, allowing them to work on social relations in Calabar and beyond that may bring rewards, such as airtime. (Gilbert 2016: 10)

Mobile phones thus function as instruments for creating a space in which to cultivate a particular identity and 'to connect with others without actually having to know them' (Gilbert 2016: 12).

It is precisely this capacity for creating intimate strangeness or strange intimacy that renders mobile phones in PNG as troubling symbols of domestic discord. I refer here to the prevalence of 'phone friends', a phenomenon found in other countries such as Morocco (Kriem 2009) and Bangladesh (Huang 2017) and among Maasai herders in Tanzania (Baird 2021). 'Phone friend' designates a relationship initiated through a randomly dialled call that might or might not evolve into a face-to-face relationship, usually but not always between a man and woman (for PNG, see Andersen 2013; Jorgensen 2014). Phone friendship has offered new opportunities for experiencing intimacy at a distance, which has in turn led proximate intimates, such as spouses, to negotiate new strategies for maintaining trust.

In her rich and definitive account, Barbara Andersen (2013) describes how several stories that bear traces of local mythology have grown up around phone friendships in PNG. These stories serve as stark warnings about the disruptive effects of mobile phones on marital relations. For example, in one story, a woman has a phone friend who sends her money to buy whatever she pleases. This woman becomes pregnant twice, each time giving birth to a snake, which she kills. When she becomes pregnant the third time, her phone friend announces that he is coming to see her. He reveals himself to be a snake and instructs her not to kill his newborn child lest he kill her entire family. The woman had reciprocated the gifts of her phone friend not with her reproductive services, but instead with the death of his children, thus erasing the chance of establishing proper affinal connections. Jorgensen (2014) likewise reports a story of a woman in Tabubil whose phone friend in East New Britain is really a snake—a tale apparently based on a traditional story of Magalim ('Bush Spirit') known throughout the region.

The fantastic nature of these cautionary tales predisposed me to wonder about the reality of phone friends in PNG, and to speculate that the tales were narrative symptoms of a moral panic. My own experience suggested otherwise. On one occasion, while waiting for a call, I answered my mobile phone despite not recognising the number. No one responded, however, to my hello. A few minutes later, I received a lascivious text message from the same number, written in Tok Pisin, expressing the sender's desire to provide me with a certain sexual service. On another occasion, I made the mistake of answering the phone after midnight, quickly disconnecting when I realised it was from an unknown number and silencing the ringer. The next morning, I noticed that the same number had called a dozen more times, perhaps because the caller was taking advantage of Digicel's dubious promotion of free airtime after 11 pm (see Chapter 3).

Further inquiries confirmed the pervasiveness and serious consequences of phone friendships. Over breakfast at a guesthouse in Goroka, I fell into conversation with a middle-aged man from the area who was working at one of the mines in PNG and had returned home for a two-week break. He asked about my research and eventually the topic of calls from unknown numbers and phone friends came up. The man told me that his own marriage broke up because of his wife's phone friend. While he was away at work, his wife, the mother of teenage children, formed a phone friendship and left him for a man who lived in the same region of the highlands from which she originally came. My interlocutor eventually remarried and started a new family with a woman from Goroka. He concluded that mobile phones were not good for a developing country like PNG, unlike developed countries where 'people get used to the system'.

This story was not the only one I heard about domestic discord due to phone friends. A university-aged researcher told me about the experience of his father and his father's young third wife, rural villagers in the eastern highlands. The father asked to borrow his wife's phone. She refused. He grabbed the phone away and discovered text messages and records of calls from another man. The woman ran away, accompanied by other women married to her husband's brothers. A large meeting eventually had to be convened, with representatives from the men's side and the women's side attending, in order to repair the rift. This researcher knew of another senior man whose phone friend became a junior wife in a polygynous household. This senior man, I was told, treated his new wife poorly, telling her that she is 'just a phone friend' to whom he has no obligation to listen.

Another university researcher, Andrea, shared her personal experiences with phone friends. On two separate occasions, she arranged with a phone friend to meet in person. Andrea travelled from the highlands to a coastal town for the rendezvous. On the first occasion, her phone friend failed to show up, leaving Andrea to feel that the relationship she thought was one thing her phone friend thought was another. An intimate had become a stranger. On the second occasion, Andrea's phone friend described what he would be wearing for their rendezvous. Andrea lied about what she would be wearing. When Andrea arrived at the arranged meeting spot, she discovered that her phone friend, despite his young sounding voice, was an old man with missing teeth (*lapun tit bruk pinis* in Tok Pisin). She ran away without revealing herself.[1] Andrea noted that this second occasion happened before the popularity of WhatsApp, which made sharing of photos easy. Previously, she said, it was just voice and 'sexting', and there was an exciting mystery about what one's phone friend looks like.

Concerns about phone friends in particular and mobile communication in general lead married partners to different strategies about how to define and respect each other's privacy. The GSMA report on low-income women indicates 'risk averse' strategies in which women effectively surrender their autonomy:

> Not having a mobile is a risk averse strategy. Some women allow their husbands to gate keep their mobile, and access it when needed in front of him. This is another risk averse strategy to prevent problems in the household. (GSMA 2014: 31)

Similarly, I have been told of cases in which newly married wives were compelled by their husbands to close their Facebook accounts. Jealousy over post-marital contact with previous romantic partners, however, runs in both directions. A young married man thus reported how his wife discovered that he was friends on Facebook with an ex-girlfriend. The wife was able to get hold of her husband's phone and gain access to his account through the already synced Facebook app, whereupon she de-friended the ex-girlfriend.

***

Not all marital arrangements with regard to mobile phones and social media accounts smack of patriarchy. One of the most active and savvy social media users that I encountered in PNG, Basil, a man in his forties with his own business and an impressive career as an athlete in international competitions, explained that he and his wife were friends on Facebook. Each has the password to the other's account, and they ask each other about the identity of new friends that are added. Basil calls his wife when he has no data and asks her to read his Facebook messages as well as his email. Basil's wife sometimes replies to emails on his behalf.

---

1   This story resembles one told by Jorgensen (2014) about a woman named Betty.

Lisa, a university tutor in her forties with three children, said that she does not check her husband's phone and he does not check hers; they respect each other's phone and trust each other. Lisa admitted, however, that her husband does not want her to answer the phone at night or in the wee hours of the morning because unknown numbers might be men with problems trying to lure her. Men have called Lisa's number before, and she has answered the phone without thinking—not in order to entertain them, she said, but because she was half asleep. One evening, a man repeatedly called and texted. Exasperated, Lisa gave the phone to her husband. When the caller heard a man's voice he finally desisted.

Jorgensen (2014: 9), who also discovered that phone friendships among his interviewees were more common than he supposed, recorded cases in which spouses maintained phone friendships 'with their partners' knowledge and acquiescence'. But many other cases of which I heard conformed more closely to the image of the clandestine relationship that inevitably breaks up a marriage. A young university researcher shared the story of how such a relationship precipitated the divorce of her parents. One night at dinner a phone rang. The sound came from the basket her father used to carry his personal belongings. Her father ignored the ringing, claiming that it was an uncle's phone. But she noticed later that the phone remained in his basket, unreturned to the uncle. It was a Digicel phone, unlike her father's regular bmobile phone. It turned out that her father's phone friend lived in a place with only Digicel reception. The man's wife had smashed four or five of these phones, but the man would get a new phone and a new SIM card in order to step outside and surreptitiously call his girlfriend.

Jorgensen (2014) claims that women initiate random calls as well as men. His claim is supported anecdotally by the assertion of one university student that all the young women at her school have phone friends from whom they receive gifts of airtime credit. Such relationships, as Andersen's (2013) stories warn, run the risk of escalating into unmanageable face-to-face confrontations. At the same time, they apparently qualify as one way in which women can participate in the informal economy that has grown up around the mobile phone (see Chapter 3).

Phone friendships furnish novel experiences of intimacy at a distance, close encounters with strangers whose true identity one might never know and for whom one's own identity can be crafted with no requirement for verisimilitude. These experiences can be exciting and entertaining, playful and titillating communication conducted by youth beyond the surveillance of elders and relatives (Andersen 2013; Jorgensen 2014; Wardlow 2018). For women in particular, especially young urban women with limited access to public venues in the evenings, these forays into a fantasy space are kept safe by their degree of anonymity and the physical distance between interlocutors.

These experiences of strange intimacy can also be stressful and upsetting. A recently married young woman became suspicious that her husband had a phone friend. Eventually she was able to use her husband's phone to call the suspected phone friend. To her surprise, the phone friend began lecturing the young wife on how to take care of her husband properly. The young wife experienced this confrontation as a disorienting reversal of strangeness and intimacy: a complete stranger had usurped the right to speak about an intimate relationship between spouses.

Intimacy at a distance can, however, take more positive forms, ones that repair rather than disrupt domestic life. The following example, although I suspect highly unusual, reveals the efficacy of mobile phones as hopeful 'affective technologies' (Wardlow 2018). Lucy is in her early fifties, a rural woman with little education who, having been diagnosed as HIV positive, was struggling to live on her own after her older brother had refused to take her in. A previous husband offered Lucy some money and an old mobile phone. Desperate, hungry and contemplating suicide, Lucy began calling the contacts saved in the phone until one woman, instead of yelling and hanging up, agreed to speak with her. This woman, Angela, responded with compassion and began sending Lucy small gifts of food, money and second-hand clothes. Lucy and Angela never met, but continued to talk by phone, thereby restoring Lucy's feelings of hope and alleviating the anxieties that Lucy thought would reduce the effectiveness of her antiretroviral medications. A friend found by chance on a random call enabled Lucy's old mobile phone to function therapeutically as an affective technology, a medium for giving and receiving care.[2]

* * *

---

2   Unfortunately, Lucy's story did not end well. Her phone, the source of Lucy's emotional sustenance, was eventually stolen: 'All those phone friends, in Port Moresby, Mt. Hagen, and other places, they would send me credit, and we talked all the time, every night, and now I don't have a phone, and I've lost all those numbers. It's terrible. I can't stop crying about it. I had the same phone for four years, and had so many numbers, so many friends. And now it's all gone' (Wardlow 2018: 49).

The potential for mobile phones to spawn domestic discord also registers in discourse around relations between parents and children (see Horst et al. 2020). Parents predictably express concern about how their children use mobile phones when the children cannot be directly monitored. Consider the example of Basil, whose way of dealing with his and his wife's use of mobile phones contrasts sharply with his way of handling his 16-year-old daughter Sarah's use of a mobile phone. Sarah has a mobile phone, but it has no internet connection. She is not on Facebook, which Basil regards as a distraction from schooling. According to Basil, Sarah only uses the phone to connect with two numbers—his and his wife's—and then only if school is dismissed early or Basil has to delay his pickup of Sarah from school. Sarah's phone has a camera, which Basil claims is useful for doing homework, and she shows all of her photos to Basil when asking to download them on one of the computers in the family home. Basil wants Sarah to earn a smartphone as the product of her own hard work, a lesson in self-sufficiency that will serve Sarah well when her father is no longer around to care for her.

Basil's highly disciplined approach to his daughter's mobile phone use and his emphasis on success at school reflect fears among many adults in PNG that mobile phones are eroding the moral and intellectual development of youth today. A 2015 newspaper article written in connection with National Book Week reported the comments of a primary school board member, Mr A Jaima, who voiced concern that schooling had changed for the worse since the 1960s and 1970s: 'It saddens me to see that many children today do not read books. They carry around mobile phones' (Post-Courier 2015d). Jaima claimed that texting was negatively affecting spelling and reading comprehension in English for current students. Analogous concerns about distracted and unproductive workers also surfaced in PNG newspaper reports of 'IT abuse' in offices—inappropriate and excessive text messaging and use of Facebook, for example.

These fears about education are in addition to those about the use of mobile phones by children to view and circulate pornography. Lisa attempted to control this possibility by allowing her two school-age children to use only 'simple phones' rather than smartphones, but she knows fully that her kids are exposed to pornography while at school. One letter writer to *The National*, Tony Sulu Yakina of Enga Province, raised the familiar worry that children are more technologically sophisticated than their elders: 'Only some parents and guardians are technology literate and can manage, monitor and regulate the usage of technological devices that view unsolicited material.' Because parents in both urban and rural areas lack the necessary knowledge, Yakina proposed that some sort of 'filter mechanism' be devised in order to screen out indecent material (National 2015d).

In Port Moresby, discussions with parents inevitably touch on a paradox. On the one hand, it is parents like Basil and Lisa who give mobile phones to their kids, mainly out of a concern for safety. Parents understandably want to stay in touch with their children in the uncertain urban environment of the nation's capital. On the other hand, parents worry about how their kids use their phones, not only to access illicit material but also to develop ego-centred networks ('contacts') that do not include the parents and potentially run afoul of kinship-based networks (Lipset 2013). Mobile phones give children the capacity for private communication that escapes the monitoring of their elders. An instrument for enhancing communication between parents and children thus paradoxically harbours the threat of ex-communication.

Who is it? Identifying Unique Individuals

The phenomenon of 'unknown numbers' caused consternation among many Papua New Guineans soon after the arrival of Digicel. Mobile users like Lisa experienced persistent calls from unidentifiable sources as harassment rather than as opportunities to make phone friends. What could be done and by whom to identify and punish mobile users who assailed other users with unwanted calls and texts? Considering how consumers, companies and state agents addressed this question offers further insight into the dynamics of freedom and constraint within the moral economy of mobile phones in PNG.

SIM card registration is one obvious way in which the state can assert its interests in regulating how citizens use their mobile phones. In Fiji, this practice was adopted in 2010 and implemented with few logistical problems and little political opposition. In PNG, by contrast, the practice was adopted but haphazardly implemented, beset by logistical problems more so than political opposition. The contrast illuminates how the relative strength of state agents affects their capacity to shape the moral economy of mobile phones. That is, the state's capacity to attach a fixed identity to a SIM card affects the extent to which mobile phone users can access services such as mobile money, on the one hand, and become targets of government surveillance, on the other.

It is helpful to begin by recalling that in PNG the problem of attaching a fixed and knowable identity to a single phone number assumes a variety of mundane forms. The high frequency with which phones are lost, stolen or destroyed in arguments results in individuals changing numbers at regular intervals. SIM cards can also expire when not used often enough only to be replaced with new numbers. It is not uncommon to accumulate multiple numbers for a single person in one's list of contacts. Constant churn in numbers can generate unintended consequences. For example, two female friends were not responding to each other's calls and texts. Each friend wondered if she had unintentionally offended the other. One friend's phone had been stolen and replaced with a phone with a new SIM card. Her friend did not recognise this new number and declined to accept calls or read texts, while her own calls and texts to her friend's old number went who knows where.

The problem of not knowing who might be attached to a particular number also takes forms that appear exotic to observers from afar. Consider again the magazine article that appeared in *New Republic* magazine under the title 'In this Papua New Guinea Village, People Use Cell Phones to Call the Dead' (Robb 2014). The article reports on ethnographic research (Telban and Vávrová 2014) that describes how Sepik villagers attempt to contact deceased relatives by using mobile phones instead of more traditional means such as dreams and trances. The challenge that these villagers face is twofold: obtaining the correct numbers of the deceased from local healers and securing an actual connection. Villagers do not despair when their calls fail because spirits of the dead can interfere with connections. In addition, villagers 'might assume the spirits aren't available. And they ring random numbers so often that occasionally they do reach someone, whose voice they attribute to a spirit' (Robb 2014).

Ideas about the possibility of contacting or being contacted by the dead via a mobile phone crop up in other contexts as well. One research assistant observed that there was a place near her home where a canoe sank and several people drowned. When boats now pass this place, passengers will turn off their mobile phones. Similarly, another research assistant reported that a spirit (*masalai*) is said to sleep underwater near the sunken *Rabaul Queen*, a passenger ferry that went down in rough seas in 2012 killing an estimated 150 people. As boats pass by, people turn off their mobile phones lest they disturb the spirit and cause the boat to capsize.

Both of these research assistants lost friends or relatives who were aboard the *Rabaul Queen*. One of them claimed that a relative of hers had recently received a call from the phone number of one of the deceased. This relative was in the shower when the phone rang but saw the missed call on her phone log identified by the name of the deceased. Then the phone rang again. The relative answered and spoke hello, but only heard silence at the other end. Our research assistant saw with her own eyes the number recorded in her relative's phone log, which she explained by speculating that the deceased's phone had been pickpocketed before boarding the ferry and now someone else was using it.

The association of mobile telephony with mistrust, as evidenced in phone friend stories about serpentine affines, is as much a function of mobility as it is of telephony. Telikom PNG cleverly capitalised on this association in print advertisements that promoted landlines or fixed connections, over which the company retains a monopoly. One humorous print ad pictured a smiling man standing in front of a gaming machine or 'pokie' and speaking on a mobile phone: 'Oh yes boss, I'm working on those figures right now …' An adjacent photo features an empty office desk with the darkened monitor of a flat screen computer, superimposed on which block letters spell out: 'Fixed Line keeps your staff at work'.

Similar clever advertisements ran on television, depicting AWOL office workers gone fishing or otherwise engaged in the pursuit of pleasure. Telikom PNG thus offered the landline as a way to tether a known identity to a known location, thereby facilitating surveillance of workers by their employers.

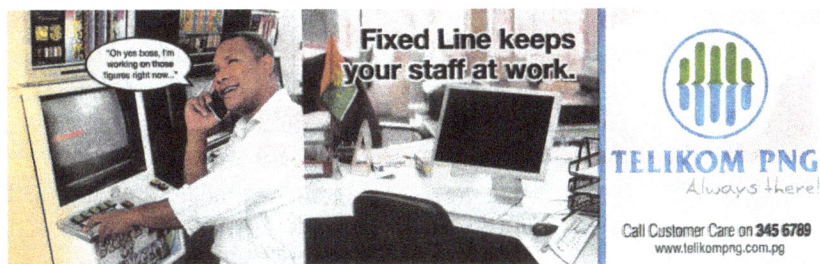

Figure 5.1. Telikom PNG newspaper ad (Fixed Line).
Source: *Post-Courier*, 10 August 2015.

Perhaps needless to say, the association of mistrust with mobile telephony extends to the social media that most Papua New Guineans access by mobile phone. A PNG Facebook friend of mine regularly complained about identity abuse with comments such as these, which invariably received enthusiastic support:

> Facebook users who use pen names and objects as their profile photos are FAKE!!!

> There are many types of scams on FB! The main one is 'males having female photos as their profile pictures'

Another PNG Facebook friend has used his social media account to inform his acquaintances about his mistrust of unknown numbers and policy for dealing with them, concluding with the observation that 'A lot of Devils have phones these days':

> Notice to family.

> If you call or text me from a number that is not registered on my phone, I won't answer. Please always ID yourself by sending an advance text.

> Plenty *Tewel igat* phone these days.

SIM Card Registration

It is exactly this issue of making identities both knowable and subject to surveillance that underwrote some of the early public calls for SIM card registration voiced in 2012. The Chief Censor of PNG, Steven Mala, worried about the security threat posed by unknown and untraceable mobile phone users: 'Our national security is under threat and a terrorist group could come into this country and blow up the [ExxonMobil] LNG [Liquefied Natural Gas] Project by using unregistered sim cards' (Post-Courier 2012d). By contrast, Police Minister Nixon Duban was concerned with more homegrown forms of terror. A relative of one of Duban's staff members was kidnapped and demands for ransom were communicated by text messages from an untraceable number. Duban proposed to petition the attorney general and prime minister about SIM card registration. According to one newspaper item expressing his concerns, Duban believed 'that many Papua New Guineans have become victims of this powerful device but perpetrators cannot be punished because they are like ghosts who cannot be traced' (Post-Courier 2012e). The mention of ghost and devils in connection with mobile phones captures well the deeply unsettling sense of alienation that comes with the strange and unwelcome intimacy of contact by unknown others.

The move toward SIM card registration gained traction in 2013 when then Finance Minister James Marape (who became prime minister in 2019) called on telecommunications companies to act in the wake of two bomb threats that were made against government departments. In July, the National Information and Communications Technology Authority (NICTA) issued a public consultation document on draft regulations for compulsory SIM card registration, and in November, then Minister for Communications and Information Technology Jim Miringtoro announced the conclusion of the public consultation process. Miringtoro also revealed that a working committee including his department, NICTA, the State Solicitor's Office and the National Intelligence Organisation had begun formulating a cybercrime policy to combat criminal activity that 'includes everything from downloading illegal music files to stealing millions of dollars from on-line bank accounts' (National Parliament of Papua New Guinea 2013).

In May 2014, NICTA released proposed regulations for SIM card registration along with a proposed PNG cybercrime policy. NICTA's public notice, which ran in newspapers, included the following statement:

> Calls by certain sections of the community that the Cybercrime Policy and the SIM card registration Regulation are intended to police social media and block criticism of government are unfounded and misguided. The registration of SIM cards is a common practice in many countries including Australia, Singapore and Fiji where it is also used to support public safety actions and to deter and trace the use of ICT [information and communications technology] services in the commission of crimes.

The statement anticipated debate about the proposals, especially the cybercrime policy proposal, that would continue right up to and beyond the release of the final policy in October 2015. Unlike the SIM card registration regulation, the cybercrime policy lacked anything in the way of a 'thorough consultation and public awareness campaign' (Galgal 2017). In April 2016, legislation for compulsory SIM card registration was adopted, and in August the government passed the Cybercrime Code Act into law.

***

The issues of SIM card registration and cybercrime policy are closely related, but I treat them here separately in order to highlight different aspects of the relations among consumers, companies and state agents. SIM card registration raises questions about the various motivations for attaching unique identities to unique phone numbers and about the varying capacities for realising this attachment. In particular, it compels us to consider a situation in which the PNG state effectively mandated that telecommunication companies carry out a task that state agents have thus far themselves been unable to accomplish: to build a national identity database. Cybercrime policy brings into focus the question of political censorship, that is, the rights of citizens to freedom of speech and the limits of governments to regulate or even suppress criticism and dissent. SIM card registration therefore directs us to company–state relations; cybercrime policy directs us to state–consumer (citizen) relations.

SIM card registration, when decoupled from cybercrime policy, garnered a significant amount of support from many Papua New Guineans. Although considerations of national security were sometimes raised, especially in the lead-up to the Asia-Pacific Economic Cooperation (APEC) meeting hosted by PNG in Port Moresby in 2018, people more often voiced concern about morally disreputable behaviour such as hoaxes and harassment, scams and cyberbullying. A 2013 letter writer thus endorsed James Marape's initial call for SIM registration: 'I think this will solve a lot of problems from marriage breakdown to prostitution and even criminal activities like car theft and break and enter' (Post-Courier 2013a). Similarly, the president of the National Council of Women in PNG, Theresa Jaindong, supported the regulation as a deterrent to the abuse of women online (ABC 2020). Even James Marape, whose department was terrorised by a bomb threat, cited fraud and antisocial behaviour as a reason for compulsory SIM registration. Marape claimed that he was victimised by an unknown man impersonating him on a mobile phone and demanding money from a pastor:

> He told the pastor that he was me and that I had approved K50,000 for his church and that the pastor should give him K3,000 before he picks up the cheque. (Post-Courier 2013b)

From the perspective of consumers, SIM card registration presented a trade-off that has become commonplace in the digital age. A certain degree of online security and protection comes at the expense of a certain degree of privacy. Different assessments of this exchange could be found in the online commentary that followed the release of NICTA's proposal in 2014, including comments from the tech-aware members of the PNG ICT Community Facebook group. On the one hand, there were supportive comments such as these:

> For SIM registration, I give a thumbs up. It's time we registered half the people who terrorize other mobile phone users and scammers who send porn pics if you send them credits …

> SIM Card registration is good for consumer because many sick head PNGns call or text illicit contents. Time to monitor and keep the house in order so people use the sim cards for intended purposes.

On the other hand, at least one contributor raised the issue of the security of personal data:

> Am concerned that the government does not have a Data Protection Instrument in place to guarantee protection of Personal Information collected via SIM card registration and hence other projects associated with Cyber crime regulations.

Other commenters expressed scepticism about the logistics of the exercise and the lack of compelling evidence that mandatory registration actually leads to a reduction in crime. In general, however, commenters rarely questioned the morality or legitimacy of the state (i.e. NICTA) in enacting a policy of SIM registration as long as its main purpose was to protect rather than to surveil citizens.

From the perspective of the telecommunications companies charged with carrying out SIM card registration, a different set of considerations came into play. In PNG, companies were required to bear the full costs of SIM registration. In the case of Digicel, with well over 90 per cent of the national mobile market, this cost was substantial. According to a senior NICTA official interviewed in May 2015, Digicel objected during the public consultation phase for NICTA's proposals to the high cost of collecting biometric information as part of the registration process. (Collection of fingerprints as well as photos was part of the initial regulations proposed by NICTA [National 2014b].) One Digicel executive explained to me that if it takes just five minutes to fill out paperwork and take a photo or capture an image of a subscriber's ID, then it would take an hour to process only a dozen people. The costs in labour time (not to mention equipment and data management costs) would be exorbitant.

SIM registration incurred other costs for Digicel. In PNG, as elsewhere (Donovan and Martin 2014), the policy led to a decline in the number of subscribers. One report claims that between 2017 and 2018, the number of mobile subscribers dropped by almost 1.3 million, lowering the penetration rate of mobile subscriptions from 48 per cent to 30 per cent (Bryden 2021: 33). Many of the lost subscribers were likely residents of rural areas without easy access to venues for SIM registration. Even though these rural residents do not spend much money on mobile phone use, they are the recipients of calls paid for by relatives working in town. That is, even the loss of subscribers who only receive calls and texts diminished the value of the network for remaining subscribers and, of course, the mobile network operator (MNO) itself.

SIM registration does bring benefits to MNOs: in particular, benefits connected with the ongoing transition from voice and SMS to data as the primary revenue stream (see Chapter 4). Without a known identity attached to a SIM card, many of the value-added possibilities of the digital economy are moot. For example, mobile banking, ecommerce and many financial inclusion initiatives require the attribution of a unique identity to a unique number (Know Your Customer) in order to prevent fraud.

Other benefits for MNOs from SIM registration are less clear in the case of PNG. The PNG legislation apparently prohibits the use of data acquired through SIM registration for the purposes of targeted marketing (see Papua New Guinea Statutory Instrument 07 of 2016, *SIM Card Registration Regulation 2016*: Section 5.6).[3] Similarly, although the inconvenience of the registration process provides a disincentive for consumers to switch providers, in markets like PNG, dominated by one company, such considerations affect relatively few mobile users. It is possible, however, that this disincentive would make it less likely for consumers to switch SIM cards as a way of avoiding repayment of credit and bundle loans (see Chapter 3).

From the perspective of the PNG state, SIM card registration presented an opportunity to reach a goal that had thus far been elusive; namely, the creation of a national identity database. The original proposal for SIM registration put out by NICTA for public consultation stipulated that the subscriber information compiled by MNOs would 'belong to

---

3   Amanda Watson (2020b) reported that the SIM registration form that she completed included questions about Digicel services such as mobile money.

the Government of PNG'. In the final legislation, however, ownership of subscriber information was put in the hands of the individual licensee that acquired the information.

The feasibility of SIM card registration is to a large extent a function of whether a country already has a national identity scheme in place (Watson 2018; GSMA 2016). Right at the start of public calls for SIM registration in 2012, Digicel CEO John Mangos urged stakeholders to consider creating a national ID system before creating a network user registry (Post-Courier 2012d). Yet by the time SIM card registration regulations were passed in April 2016, little progress had been made in this regard. Two years later, the National Identification project (NID) initiated in 2015 had registered only a small fraction of the population: 'In the years since, only around 500,000 of PNG's approximately 8 million people have been registered, the card printing machines have broken down and the bureaucrat who administered the scheme has been accused of stealing and mismanaging millions of kina' (Harriman 2018). In addition, a 'syndicate style operation' selling counterfeit NID cards was uncovered. It was therefore reasonable that, one week before the SIM registration legislation was approved, a Digicel executive privately speculated that once the company had amassed its subscriber information, the database might be sold to the government in lieu of the government ever completing its own undertaking.

For the PNG state, just as for the MNOs, compulsory SIM registration entailed both costs and benefits. Some of the putative benefits, such as the development of ecommerce and e-government, were aspirations rather than directly realisable outcomes of SIM registration. And one of the most cited benefits of SIM card registration—an increase in security against terrorism and a decrease in crime—lacks persuasive evidence (Gow and Parisi 2008; Watson 2020b; GSMA 2016; Donovan and Martin 2014). By contrast, the costs of the regulations were clear and immediate: a large number of people would be excluded from the network. These people were likely to be remote rural residents who faced difficulty in navigating the registration process, people for whom mobile phones are lifelines in times of natural disaster or health emergencies. It was precisely this cost that delayed implementation of the regulations more than once (see Watson 2020b for details), as NICTA developed provisions for giving rural residents more time to register. In one instance, the delay was prompted by a lawsuit brought by a member of parliament on behalf of his rural constituents; in another instance, the delay was authorised by the communications minister in response to a devastating earthquake in the highlands regions.

The final deadline for registration passed on 30 September 2020, more than four years after the regulations were adopted. The delays in getting to this deadline bring into view some of the stakes involved in attaching unique identities to unique mobile phone numbers—that is, in answering the nagging question 'Who is it?' For some users, registration presented a moral dilemma, a choice between opting out of a communications network that has become embedded in the everyday social lives of many Papua New Guineans or exposing oneself to the scrutiny of state agencies (or indeed less knowable transnational entities; see Jorgensen 2018). Fears of policing are real, regardless of how dim the prospects are for an efficient and extensive Orwellian or Foucauldian surveillance apparatus in PNG. Hence the comments of one man waiting in a long line to register his SIM card in Goroka: 'I came to register my SIM but I'm also scared that my whereabouts will always be known by someone or hear what I say over the mobile phone' (Post-Courier 2017b).

For companies—Digicel, in particular—the mandate to register SIM cards imposed a considerable cost. But, regardless of how remote the likelihood of a vibrant digital economy in PNG appears, there is no doubt that corporate executives like bmobile CEO Anthony Pakakota speak as if their company's 'future is in the space of ecommerce' (Chai 2020). SIM registration is thus one challenging part of a larger effort to encourage and enable subscribers to use their mobile phones to transfer funds (mobile banking) and buy goods and services (e.g. to pay for electricity). In this regard, companies join the state- and non-government organisation–sponsored international project of moral improvement entailed in 'banking the unbanked' (see Peebles 2014; Chapter 6).

For the PNG state, SIM registration was also perceived to be a prerequisite for other initiatives intended to improve the lives of citizens, such as enhanced protection against cybercrime and efficient delivery of government services. As already mentioned, however, compulsory registration prioritised an imagined future over an actual present inasmuch as the regulation resulted in the deactivation of an estimated million-plus SIM cards. Similar negative effects on connectivity—especially for the poor—have been observed in other developing countries (Gillwald 2015a, 2015b; Donovan and Martin 2014), raising doubts about the moral legitimacy of the state's actions.

The numerous delays in implementing registration and the exclusion of unregistered subscribers from the network prompt the further question of whether SIM card registration in PNG, at least at the moment, was worth it (see Watson 2020b). It is perhaps the case that the policy along with its cousin the Cybercrime Act were adopted in a process of 'international normalisation' (Gillwald 2015a) or 'institutional isomorphism' (Dimaggio and Powell 1983, cited in Donovan and Martin 2014). In other words, the adoption of the policies might best be explained as a form of mimicry in which organisations model themselves on what organisations in other countries are doing (as NICTA's public notice in 2014 clearly implied):

> In the case of SIM registration, although states have little evidence of the advantages, transformations wrought by the growth of largely anonymous mobile communication create a situation of unpredictability where imitation and mimicry is likely to occur. (Donovan and Martin 2014)

This sort of mimicry is more obvious when there is a perceived need to harmonise laws in one country with laws in other jurisdictions, as in the case of cybercrime regulations, to which I now turn.

Cybercrime Laws and Online Censorship

The swift uptake of smartphones through which most people access the internet has raised concerns about freedom of political expression. In PNG, where social media and blogs quickly became vehicles for criticism of government policy and individual politicians, these concerns accompanied the introduction of cybercrime legislation. This legislation and its implications for the expression of political speech has tempered hopeful claims about the potential for mobile phones to enhance transparent and democratic governance.

Although closely related to SIM card registration regulations, cybercrime policy addresses a problem different from that of attaching unique identities to unique mobile phone numbers; namely, the use and abuse of ICT services and communications devices for illegal purposes (see Section 266 of the *National Information and Communications Technology Act 2009* [the NICT ACT]). A very large range of offences fall under the heading of cybercrime, including cyberbullying, hacking, sending indecent materials, spreading false information, spamming, electronic fraud, defamation and sedition. The *Cybercrime Code Act 2016* is far-reaching in its coverage, presenting real challenges in balancing 'the desire for safety and security with the need for free speech and freedom of association' (Oxford Business Group 2016).

It is fair to say that most Papua New Guineans welcomed and supported regulation of online harassment and, in particular, the distribution of pornography. This was especially true after the front page of the 17 January 2017 *PNG Post-Courier* newspaper shouted: 'PNG Tops World in "Porn" Search'. As Watson (2017) noted in her analysis of the report:

> As PNG prides itself on being a Christian country with strong traditional cultures and values, coupled with tough laws banning importation of pornographic magazines and movies, the headline has produced consternation.

Although Watson's conclusion that the newspaper report was 'alarmist' is persuasive, concerns about consumption of pornography regularly appeared in letters to newspaper editors. An ad in the 16 March 2015 *Post-Courier* invited readers to participate in a text poll by texting yes or no to the following question: 'Do you think PNG has a pornography problem?' Similarly, my conversations with staff members of the Office of Censorship in 2015 also made it clear that research into the use of mobile phones by school children was a top priority (see Morofa 2014).

Critics of the new cybercrime laws focused instead on potential use of the laws in suppressing criticism of government policy and government officials, especially on social media platforms. In an interview with Radio New Zealand, Martyn Namorong, a prominent PNG netizen (Capey 2013), expressed his concerns about provisions in the cybercrime law regarding 'cyber unrest':

> You know, some of my blog posts are very provocative and could be interpreted as trying to undermine the state. I mean, is that sedition? It's very unclear whether some of the criticism of government could be interpreted as undermining the state. For instance, you use social media to talk about protesting or organising people to protest, is that undermining the state or undermining the government? (Radio New Zealand 2016)

Former PNG prime minister Sir Mekere Morauta was more forthright in his denunciation of what he perceived as the misuse of government regulation:

> The Prime Minister's [Peter O'Neill] use of the NICTA Act to clamp down on freedom of speech and the media is an unparalleled abuse of power. The nation would seem to be moving step by step towards becoming a dictatorship. (Pacific Media Centre 2015)

NICTA, for its part, publicly responded by reminding social media users of Section 266 of the *NICT Act 2009*, which stipulates the punishments for:

> a person who, by means of an ICT service:

> Sends any content or communication that the person knows is offensive or of an indecent, obscene or menacing character, or;

> 1. For the purpose of causing annoyance, inconvenience or needless anxiety to another person –

> Sends any content or communication, that he/she 1. knows to be false, or;

> 2. persistently makes use of that ICT service with that intended purpose, is guilty of an offense.

The introduction of mobile broadband in PNG lifted hopes that digital communications would lend greater transparency to government operations and aid in the fight against corruption (Logan 2012; Rooney 2012).[4] As early as 2009, with the launch of the Namorong Report (Capey 2013), Facebook groups and blogs reported and circulated news about alleged government malfeasance and failure of service delivery (see Logan and Suwamaru 2017).[5] The rapid spread of smartphones following the introduction of Digicel Broadband in 2011 increased the velocity of circulation and made news more accessible to more people. Social media in PNG abounds with posts and tweets that, although not always accurate, criticise both government policy and individual politicians. Even some users of basic handsets—government officers working in provincial and district treasuries—can anonymously report cases of corruption through a text messaging service set up by the PNG Department of Finance with the support of the Australian Government and the United Nations Development Programme (Watson and Wiltshire 2016).

---

4   Basic handsets had already been put to this purpose. In 2010, Peter Aitsi, former PNG head of the anti-corruption organisation Transparency International, coordinated a protest against government efforts to undermine the Ombudsman Commission by using mobile phones to distribute information to community groups and NGOs. Aitsi claimed that 'Mobiles democratize information flow in PNG. People in the village who can't rub two kina [about 80c] together can now communicate' (Hendrie 2011).

5   See, for example, the following Facebook groups: Sharp Talk, created 2011 (www.facebook.com/groups/Sharptalk/); PNG News, created 2012 (www.facebook.com/groups/326819464091972/); Act Now!, created 2009 (www.facebook.com/ActNowpng1). See Rooney (2012) for more examples.

There is no doubt that online activism caused concern among government officials. Logan (2012) observes that as early as 2010 the anti-corruption efforts of internet users had elicited attempts at government censorship:

> Bloggers circulated leaked reports of a major corruption enquiry that directly implicated senior political figures, and which the government was unwilling to release. When the report started surfacing on blogs, the government delivered a writ to the country's major internet service provider (ISP), ordering it to block blogs which hosted the report.

Although this heavy-handed attempt at media control proved unsuccessful, it testified to the disruptive effects of emerging uses of ICT by ordinary citizens.

The use of social media to organise, witness and document political protest first occurred in a big way in 2012 when for a brief, strange and tense period two different men claimed to be the prime minister of PNG (see Cave 2012: 12; Cranston 2013; Logan and Suwamaru 2017). In 2016, a public protest in Port Moresby that led to a tragic confrontation with police and the shooting of several university students was well documented on Facebook and other social media platforms. International news reports now recirculate images grabbed from social media in PNG, thus amplifying their effects as testimony to current events in a country where few foreign correspondents reside. The Twitter handle of a PNG woman I had met the year before appeared on a video of the 2016 shooting that accompanied a *Guardian* article describing the incident. This woman, an attorney educated in Australia who works and resides in Port Moresby, regularly uses her Twitter account and two iPhones to comment provocatively on local politics.

There is equally no doubt, however, that ordinary citizens engage in less salubrious forms of online critique. Consider, for example, the media scandal that erupted in 2015 over the online circulation of a 'nude selfie', alleged to be an image of Fred Konga, who at the time was executive chairman of the Border Development Authority, standing naked before a mirror with mobile phone in hand. The image was posted to a blog called 'Caught in the Act', reposted on other blogs, circulated through Facebook and email, and written about in *The National* newspaper. Konga, who denied taking the selfie and claimed that the image was fake, brought a defamation suit against several people who published the photograph, the journalist who wrote articles about the image and *The National* newspaper.

Konga also called for tighter regulation of media and speedier implementation of SIM card registration. His response, in turn, elicited further publicity and criticism on social media, including a post on PNGBLOGS from We Are Anonymous PNG, 'a loose knit global group of concerned citizens who fight for justice and fight against corruption' (PNGBLOGS 2015). The post declared that the alleged selfie was 'a display of a PNG government bureaucrat behaving like a child not befitting the office he occupies'. We Are Anonymous PNG pledged to continue to 'use cyberspace to take power back from the corrupt and put it once again into the hands of law abiding citizens' (PNGBLOGS 2015).[6]

Online critique of the kind exemplified by the case of the nude selfie often formats political expression as a genre of resigned, often ironic, commentary. For example, in July 2016 a photo of the Speaker of Parliament, distracted by his smartphone in the middle of an unsuccessful vote of no confidence against Prime Minister Peter O'Neill, was posted to one of the popular Facebook news groups. It garnered more than 800 comments responding to the question 'What was he doin?' The droll responses included: checking his balance, texting the prime minister, phone banking, watching porn, and playing Pokémon Go. Finally, after four days, one commenter called for an end to the commenting and suggested that people stop wasting their time and wait until the next election when they could vote for a change of government.

This sort of frustrated intervention, which calls attention to the uncertain relationship between online and offline action, is not uncommon on news groups. It was a regular feature of Digicel Complaints Group (DCG), a public Facebook forum for voicing grievances against Digicel (see Chapter 4; Foster 2020). Sooner or later, after a long string of complaints about missing credits, someone would remind the group not only that Digicel is not listening, but also that Digicel is actually making money from all these complaints. And this comment is inevitably followed by the suggestion that everyone quit spending their time complaining about Digicel and simply switch to bmobile. Such is the slippery slope of consumer-citizenship, whereby collective political initiatives devolve into individual market decisions (or yet one more moment in the 'politics of resignation'; see Benson and Kirsch 2010).

---

6   The *PNG Post-Courier* reported in 2017 that Fred Konga was killed in an 'execution style' shooting in Port Moresby (George 2017). Konga had resigned his position as CEO of the PNG Border Development Authority in order to contest the election for the Jiwaka Province regional seat.

Logan (2012) has suggested that when the increased transparency afforded by social media is coupled with a continued lack of government accountability, the net result is cynicism. Thus, for example, social media was useful during the 2017 national elections in documenting the failure of officials to turn up at polling stations, the inability of voters to find their names on the electoral rolls and the practice of obviously underage 'voters' casting ballots. Indeed, the inability of state agents to fix and record the identity of PNG citizens was highlighted as concerns over inaccuracies in the electoral rolls undermined the legitimacy of the voting. All these acts of witnessing were not enough, however, to prevent the return of the same government and the same politicians so regularly and roundly decried online in forums such as PNGBLOGS and DCG. As if ineluctably, the same failure of officials to make the electoral rolls accurate and up to date marred the national elections in 2022.

Despite the emergence of a species of resigned political discourse à la DCG, it would be wrong to make an invidious distinction between a hermetically sealed online pseudo-politics and a 'real' on-the-ground politics. Indeed, Bryan Kramer, who helped launch DCG, was able to use social media as a platform upon which to run successfully for election to parliament in 2017 as a representative from the town of Madang. His Kramer Report on Facebook claimed 100,000 followers in July 2018. Nor did his election as member of parliament precipitate a turn away from strategic consumer-citizenship. In 2018, Kramer used social media to mobilise a public around the boycott of PNG's two major newspapers on the grounds that the country's print media had failed to report truthfully and effectively on government malfeasance. Kramer has continued to use social media as a vehicle for challenging government policies, and he was both an important online witness to and engaged participant in the opposition that led to the resignation of Prime Minister O'Neill in May 2019.

In short, politics as usual has become enmeshed in and reconfigured by the changing mediascape in PNG. Yet the state's capacity to enforce its 2016 cybercrime legislation remains unclear. In August 2016, the Minister for Communications and Information Technology, in an announcement on the Facebook page of the ruling People's National Congress party, 'called for civil legal action to be taken against online news sources who publish false and misleading information that costs millions of Kina in lost national investment, income and employment'.[7] The minister named offenders that included EMTV (a PNG television broadcaster) and the Australian Broadcasting Corporation (ABC), but no lawsuits were actually brought.

In 2018, PNG attracted rare international media coverage when Sam Basil, then communications minister, announced a one-month ban on Facebook. According to a report in *The Guardian* (Roy 2018), Basil told the *Papua New Guinea Post-Courier* newspaper:

> The time will allow information to be collected to identify users that hide behind fake accounts, users that upload pornographic images, users that post false and misleading information on Facebook to be filtered and removed … This will allow genuine people with real identities to use the social network responsibly.

Both Basil and Prime Minister O'Neill subsequently denied any government plans to implement such a ban. However, in May 2019, as momentum built toward a vote of no confidence against him, O'Neill announced a new initiative to regulate social media and to prevent the circulation of 'fake news'. The prime minister asserted: 'It is destroying our people and destroying our society. We've lived in peace and harmony for thousands of years without social media' (Elapa 2019). O'Neill focused his criticism on 'multi-billion dollar foreign companies' and on Facebook, in particular, for its role in the promulgation of false information.

The prime minister's announcement was met with immediate criticism from politicians and journalists. Governor Gary Juffa's rebuttal on Facebook was representative:

> We are all subject to abuse and gossip and rhetoric and though we may be upset or hurt by it we need to ensure we protect the rights of our people to express themselves as that's an essential aspect of [what] democracy is about—freedom of speech. If we have issues we can report it to the Police or take civil legal action.[8]

Sylvester Gawi (2019) likewise blogged:

> The fact is you can't control platforms where information is circulated, attempts to do such undermines the role of democracy and freedom that is enshrined under the constitution of our country.

---

7 See: www.facebook.com/peoplesnationalcongress/posts/-online-news-reporters-should-face-legal-action-for-misreporting-authorised-by-t/575162216020378/, accessed 6 December 2022.
8 See: www.facebook.com/juffa/posts/10155910867102134, accessed 6 December 2022.

In the event, nothing came of the prime minister's plans and his resignation ultimately rendered the whole matter moot, at least for the time being.

***

It remains uncertain what effect, if any, SIM card registration and cybercrime laws have had on public discourse several years after their enactment. What seems clearer is perhaps a corollary to Logan's (2012) proposition that transparency without accountability breeds cynicism. That is, if the ineffectiveness of state agents to eliminate corruption and curtail the abuse of political office promotes resignation, then the ineffectiveness of these same state agents to enforce censorship and surveil users enables and may even promote continued critique. Not only digital activists such as Martyn Namorong but also elected officials such as Bryan Kramer and Gary Juffa have resisted recurrent efforts to govern the infrastructural assemblage in ways that suppress criticism of government.[9] The unfolding of these efforts and counter-efforts—inevitable given the felt need in PNG as elsewhere to deal with the circulation of hate speech and disinformation through social media (see Kant 2022)—will reshape the dynamics of freedom and constraint within the moral economy of mobile phones.

---

9   For accounts of digital activism under conditions of tight media restrictions in other Pacific Islands countries see Titifanue et al. (2016) and Brimacombe et al. (2018).

6

Connecting the Unconnected: Corporate Social Responsibility and Post-Political Governance

Introduction: Corporations as Caring and Responsible Persons

Renewed interest in corporations on the part of anthropologists has directed attention to corporate social responsibility (CSR), the various strategies and means that corporations use to represent and enact their commitments to more than the bottom line (see e.g. Welker 2014; Rajak 2011b; Dolan and Rajak 2016). The flourishing CSR industry—with its own companies, experts and events—is increasingly the subject of ethnographic inquiry (Rajak 2011a; Garsten 2010; Conley and Williams 2005). Rajak (2011a: 10), for example, treats CSR conferences held in London as part of the 'ritualized and performative dynamics of CSR', 'theatres of virtue' in which participants speak of 'win-win solutions' in which multinational corporations do well by doing good. The repeated use of key terms—'transparency', 'accountability', 'partnership'—effectively brings into being an exclusive 'discursive coalition or community' (Garsten 2004; Garsten 2010: 60). Alternative visions and critical vocabularies get filtered out: high conference fees prevent many activists from even attending let alone speaking. Rather than see these events as cynical exercises in public relations, Rajak (2011a) understands CSR conferences as powerful devices for consolidating a consensus about how global economic development ought to unfold and who should lead the way forward.

Ethnographic inquiry of this sort is linked to broader conceptual questions about the shape of neoliberal capitalism. As Ronen Shamir (2008: 1) observes, 'contemporary tendencies to economize public domains and methods of government also dialectically produce tendencies to moralize markets in general and business enterprises in particular' (see also Shamir 2010). In other words, the rise of CSR and business ethics—what Shamir terms the 'moralization of markets'—goes hand in hand with the economisation of state and civil society institutions through privatisation and deregulation. What does this dialectic or two-sided process look like in practice, specifically, as a feature of the moral economy of mobile phones in Papua New Guinea (PNG)?

In a world of 'downsized states and unconstrained global corporations' (Ferguson 2005: 378), it is hardly anomalous for companies to operate outside the purview of the nation-state, securing private capital with private armies. But the moralisation of markets just as often assumes the superficially benign form of 'post-political governance', a kind of global market regulation in which the traditional role of government gives way to 'voluntary regulatory arrangements, soft law and moral regulatory frameworks such as codes of conduct, standards for corporate social responsibility (CSR) and the like' (Garsten and Jacobsson 2007: 145). The world of post-political governance is full of 'partnerships' between corporations, on the one hand, and civil society organisations, non-government organisations (NGOs) and state agencies, on the other. That is, states are merely one type of actor sharing rule-making authority with other actors (Garsten and Jacobsson 2007: 147). These partnerships purport to deliver goods and services in a manner that benefits all parties—as well as 'the people'—in a collaborative manner that, moreover, eliminates any political conflicts between the different interests of states seeking social accountability and businesses seeking profits. In some instances, the boundaries between states and corporations are difficult to distinguish (Shever 2012; Welker 2014), thus diminishing the capacity of 'the people' to make claims as citizens rather than as consumers or clients.

Corporate strategies that target people living at the 'bottom of the pyramid' (Prahalad 2010) exemplify well how post-political governance operates. These development initiatives, which have caught the critical eye of anthropologists, are advertised as providing both affordable solutions to the problems of people living in poverty—poor nutrition or sanitation, for example—and a desirable mass market to the companies selling these solutions. Nestlé's Maggi brand instant noodles will reduce hunger in PNG (Errington et al. 2012); Lever Brother's Lifebuoy brand soap will remedy bacterial infections in India (Cross and Street 2009). Caring multinational corporations vend 'social goods' and dignify developing world inhabitants by addressing them as legitimate consumers, while at the same time accomplishing business goals. Does everybody win? Not necessarily, if one imagines the goals of development to include jobs that pay wages sufficient to buy nutritious food or an infrastructure that makes access to clean water widely available to people on the basis of their status as citizens rather than consumers. Indeed, many of these initiatives effectively compel (or 'empower') people living in poverty to take personal responsibility for their situations by changing their own behaviour rather than changing oppressive relations of political power and authority (Shever 2010).

It is difficult not to critique the contemporary discourse and practice of CSR as a tool for turning adversaries into partners, for deflating and deflecting conflict by holding out the promise of consensus (Foster 2014a). Marketing managers and public relations officers strive to represent corporations as not only caring, but also rational and open to dialogue with critics. Accordingly, individuals and groups, often with far fewer resources than those available to a global corporation, who resort to noisy protest are represented as uncivil and hostile (see Foster 2014b). Understanding and challenging how CSR underwrites corporate responses to critique is thus an urgent task for any critical anthropology of corporations. Benson and Kirsch (2010) and Kirsch (2014) have made important contributions in this regard by documenting the remarkably standard ways in which 'harm industries' such as mineral extraction and tobacco manufacturing attempt alternately to deny, deflect and absorb criticism, often in an effort to pre-empt legal regulations.

This chapter documents some of the ways in which Digicel, through the Digicel Foundation in PNG, discharges obligations commonly regarded as functions of the state, such as building schools and health clinics. It asks how CSR constructs telecom companies as morally responsible citizens while at the same time making markets for the goods and services that these companies sell. These goods and services include life insurance policies and, in particular, access to formal banking proffered in the name of financial inclusion.

Digicel's business is of course not comparable to that of a 'harm industry' like mineral extraction or tobacco manufacturing. Its manifest success in making telecommunications services available and affordable to a large number of Papua New Guineans deserves acknowledgement. By the same token, Digicel's status as a for-profit corporation subject to the demands of its investors also deserves acknowledgement. Digicel never proclaimed its mission as trying to alleviate poverty, despite the fact that its creation of a market of prepaid subscribers employed a key bottom-of-the-pyramid strategy—namely, offering airtime top-up in small inexpensive units. Digicel's mission was to deliver telecommunications services to consumers at a rate that would ensure net revenue for the company.

How, then, did Digicel's commitment to CSR shape its position in the moral economy of mobile phones and, specifically, its relations with PNG state agencies? How did Digicel participate in partnerships of the sort characteristic of post-political governance? How did these partnerships and other corporate-led efforts to do social good follow, or not, the precepts of a Melanesian gift economy?

Foundation Work

The work of the Digicel Foundation in PNG illustrates plainly how corporations can act as states, for much of the foundation's work in delivering social services resembles what a citizen might reasonably expect of a modern state. Digicel operates foundations in only four of its markets: Trinidad and Tobago, and its three largest markets, Jamaica, Haiti and PNG.[1] In PNG, the foundation operates across a larger range of target areas than in other countries: education, health, community building (e.g. teaching business skills to unemployed youth), special needs (e.g. helping persons living with disabilities) and addressing violence (e.g. empowering women) (see Digicel Foundation 2018/19). In PNG, moreover, the foundation has been closely identified with the company. Digicel PNG officers, including the CEO, have served as foundation board members, and in 2015 the foundation shared office space with the company. Digicel staff were also regularly encouraged to volunteer for activities sponsored by the foundation.

---

1   It was reported that Telstra planned to continue operating the Digicel Foundation after the acquisition of Digicel Pacific in 2022 (National 2021).

The foundation was established in 2008, one year after Digicel arrived in PNG. It was funded wholly by Digicel and although a separate entity, its finances 'go through the business' to ensure sound management (Watson and Mahuru 2017). In 2015, then foundation head Beatrice Mahuru told me that the foundation's budget was about USD4.5 million (personal communication, 2 April 2015), a sign of Digicel Group Chairman Denis O'Brien's strong belief that as Digicel grows so must the communities in which the company operates. The foundation's staff was small, about a dozen employees, and 15 per cent of the budget went toward administration (staffing and travel) (see Watson and Mahuru 2017). In its first 10 years the foundation completed 433 projects (362 infrastructure projects and 71 social programs), investing a total of USD29.19 million (Digicel Foundation 2018/19).

In a 2017 interview (Watson and Mahuru 2017), Mahuru indicated that the foundation had begun 'to work more collaboratively' with the business by promoting goodwill in response to negative feedback that the company receives, especially on social media, regarding the cost of service:

> We grow goodwill through our development impact projects and lay the foundation on which business can follow. Last year, we probably put 40 per cent of our investment into promoting goodwill and the rest was responding to [community] applications received.

Mahuru's comment might be interpreted as evidence of how companies regard CSR as a means for doing well by doing good. But it might equally be interpreted by cynics and critics as evidence of how CSR is no more than a self-serving exercise in marketing and brand enhancement. In either case, it is apparent that the mission of the foundation was closely aligned with that of the business.

The foundation's emphasis on partnerships recalls the similar strategies used by the company to expand its cell tower network in PNG (see Chapter 1). For example, in 2014 the foundation accepted a cheque for PGK75,000 from member of parliament (MP) Bob Dadae for the construction of a double classroom in his remote Kabwum District (Morobe Province) (Loop PNG 2014). The MP's contribution would cover approximately half of the cost of the new classroom facility. Mahuru observed in 2015 (personal communication, 2 April 2015) that such public–private partnerships were advocated for by Prime Minister Peter O'Neill even before the creation of a relevant regulatory framework in 2014 (the *Public Private Partnership Act 2014*) for the procurement and delivery of infrastructure projects and services.

The close alignment of interests and strategies between the company and the foundation was visible in the agreement struck in 2018 with the PNG Sustainable Development Program (PNGSDP). PNGSDP allocated USD32.5 million for Digicel to upgrade the company's 78 towers in Western Province for 4G services and to build an additional 19 new towers to be owned by PNGSDP (Business Wire 2018). At the same time, PNGSDP funded a PGK20 million program to improve facilities in 35 schools. The Digicel Foundation was paid PGK5 million to develop new classrooms in eight Western Province schools during the first phase of the program (Post-Courier 2018b). The foundation, moreover, apparently entered into partnerships with foreign governments. In 2017, it received grant funding from the Australian Department of Foreign Affairs and Trade 'to implement education projects' in the Autonomous Region of Bougainville (Watson and Mahuru 2017). These arrangements effectively cast the foundation in the role that companies play in carrying out Universal Access and Services projects whereby the state (i.e. the National Information and Communications Technology Authority, or NICTA) allocates tax revenue collected from telecoms to extend ICT (information and communications technology) infrastructure in underserved parts of the country.

Reciprocally, the foundation entered into partnerships in which it plays the main role of donor or funder. For example, the foundation partnered with Cheshire Disability Services to conduct assessments and training in 'marginalized communities' in the National Capital District and to provide 'physio-therapy services for people living with disabilities' (Digicel Foundation 2018/19: 28). Besides partnering with community organisations, churches and NGOs, the foundation also commissioned businesses to carry out projects. In 2018, for example, the foundation signed an agreement with Total Energy and Petroleum Ltd to build classrooms and rural health aid posts in Gulf Province (Digicel Foundation 2018/19).

According to the foundation's 2018/19 annual report, the 'School Infrastructure Program is the flagship program of the Digicel PNG Foundation'. Through this signature program, which accounts for 60 per cent of the foundation budget (Watson and Mahuru 2017), the foundation has built hundreds of lightweight-steel, low-maintenance classrooms in communities across the country. Indeed, the foundation prefers to spread its projects across as many different locations as possible, rural and urban, rather than concentrate them in one region (see National 2013). The classrooms, which are painted red and white and affixed with a Digicel Foundation sign, come with office space for teachers, toilets and a water tank and solar lighting. These projects are represented in recognisably developmentalist terms as comprehensive materialisations of modernity:

> Damp, dark, bush material classrooms were removed and in place of them, well ventilated, steel classrooms with solar, ramp access, desks, teachers office space and water sanitation component consisting of toilets, bucket showers and 9,000L water tank were given. (Digicel Foundation 2018/19: 15)

The school infrastructure program is motivated by a particular moral vision of responsible self-improvement, a vision consistent with the claim that 'Digicel Foundations are run like a business and are not a cheque writing/ handover charity organisation' (Digicel Foundation 2018/19: 48). Mahuru noted that the foundation expects and enforces accountability—not always easy when working with politicians (personal communication, 2 April 2015). Applicants to the foundation must demonstrate their viability as well as meet certain criteria. Potential sites are inspected beforehand and rejected if there is evidence, for instance, that the building would not be taken care of or that disagreements about land ownership would present problems. The foundation's emphasis on self-help—phrases such as 'human resource development' and 'community based capacity building' recur in foundation press releases—is apparent in a statement from 2011 that defined the requirements applicants needed to meet in order to secure a Community Learning Centre (CLC) for their community:

> In order for a community to receive a CLC building they must apply to the Digicel Foundation and demonstrate that they have strong community leadership, motivation, ownership and responsibility for all members in their community. The Digicel Foundation works with those communities that are active in helping themselves with the little resources they have. (Digicel Foundation, '21 Community Learning Centres (CLCs) in Port Moresby, Mt Hagen, Goroka and Lae compete for prizes worth K18,000', statement, 26 April 2011)

Similarly, the foundation's signature Men of Honour public awareness campaign against domestic violence promotes the stories of men who have taken responsibility for improving their lives and the lives of others. The campaign recruits these men to run workshops and serve as models of a morally desirable kind of masculinity.

The moral vision of self-improvement and self-sufficiency projected by the foundation's infrastructure program might not be shared unequivocally by all recipients of classroom buildings. On two different visits to foundation classroom projects, one in Central Province and the other in Eastern Highlands Province, I heard evidence of a recognisably Melanesian gift economy from the perspective of which the foundation's actions were wanting. At one site, for example, the head teacher felt that the classroom was incomplete without a computer in the office or books for the students. This teacher and his colleagues were proud that their school building was well cared for, and they noted that Digicel Foundation inspectors who periodically visit without advance notice have complimented them on this care. The teachers felt that their care for the building (one of the teachers sleeps in an unused classroom to keep an eye on the building at night) should elicit further consideration from the foundation. That is, the teachers imagined the gift of the classroom as the beginning of a long-term relationship with the foundation rather than a one-off donation or a completed transaction.

At the other site, the head teacher similarly reasoned that looking after the 'property' and receiving the approval of foundation inspectors merited and augured additional support—maybe even another classroom. This teacher observed that his community not only takes care of the building, but also contributed materially to its initial construction. Sand was purchased and transported to the site for making the cement used to secure the posts, and the foundation's contractors were fed during the construction work. A feast, moreover, was held to launch the building. A large pig and a cow were cooked and distributed along with gifts to the Digicel staff and a collective gift of about 6,000 kina to the school. In other words, the community had established an open-ended reciprocal relationship with the foundation, a gift relationship with an indefinite future.

While the Digicel Foundation imagines itself as something other than a 'handover charity organisation', it nevertheless dictates the nature and timing of its gifts—precisely the feature that distinguishes charitable donations from gift-giving as conventionally understood by Melanesians and famously articulated by Marcel Mauss (1923–24). After all, as one of the head teachers observed, supplying classrooms does not always address the infrastructure problem that besets many rural communities in PNG—namely, the lack of housing for teachers. In addition, the close brand identification between the company and the foundation—conveyed through shared corporate logos and colours—can create confusion and ill will among the general public. During the launch of the classroom at the highlands site, the head teacher reported, Digicel sold inexpensive handsets. The head teacher himself bought two handsets at 35 kina each and later resold them at a profit. Any ideal-typical distinction between gifts and commodities (Gregory 1982) was thus put at risk, and the old question of whether CSR is simply another marketing strategy was posed again. In 2015, Digicel announced on its Facebook page that the company had responded to the news that a family in Port Moresby had lost its home in an early morning fire. The company donated items of branded merchandise that included T shirts, drawstring bags, water bottles, backpacks and umbrellas. One commenter complained:

> Honestly, for a company that is making so much to donate a bunch of promotional items is a slap in the face … . if you really want to help then do so, *inap lo giaminim ol man!!!* ['stop lying to people!!!']

My point here is not to bash either the Digicel Foundation in particular or CSR in general; to do so in the larger context of a national economy warped by harmful resource extraction and endemic official corruption would be peevish. My aim is instead to view the foundation in light of the moral economy of mobile phones as an actor that takes on many of the obligations associated with modern states, such as the provision of critical infrastructure. But the foundation is not a state agency; it was practically an arm of Digicel PNG's business. To the extent that the foundation is identified with the company, its activities define expectations on the part of consumers (not citizens)—expectations of rewards for continued loyalty comparable to those of a classic big man's followers (Sahlins 1963; Clark 1997) or indeed a regular user of Digicel's prepaid services (Chapter 3). When these expectations are met, goodwill results; when these expectations are not met, goodwill can be impaired. Because goodwill signifies an economic good, an intangible asset on the company's balance sheets, and a moral attitude, a conviction on the part of consumers, it is a highly contested and unstable object, produced at the interface between companies and consumers in the moral economy of mobile phones. In PNG, this interface sometimes brings into uneasy contact different and incompatible ideas of what morally legitimate persons owe each other. It is a space where, in the absence of the state caring for its citizens, consumers and companies make competing claims on each other.

The Quest for Financial Inclusion

Evidence of companies and other non-state actors delivering goods and services to the people of PNG in the name of national development can also be observed in the area of digital financial services (DFS). The terms 'financial inclusion' and 'banking the unbanked' evoke efforts to help people imagined as marginalised (excluded) from mainstream economic life or otherwise neglected at the 'bottom of the pyramid' (Prahalad 2010). In PNG, as elsewhere, mobile phone companies are important actors in these efforts inasmuch as almost all initiatives presume that new forms of DFS such as mobile banking will be accessed through mobile phones—both basic phones and smartphones. What do these efforts look like in PNG and what role do mobile network operators (MNOs) play? What are the prospects and problems involved in designating and using mobile phones as instruments for enhancing economic and material wellbeing?

The prospects of using mobile money (e.g. electronic money transfers and payments) and mobile financial services (e.g. banking and borrowing) generate lots of excitement and enthusiasm in PNG for a variety of compelling reasons. The quick uptake of Digicel's Credit Me/Credit U service (Chapter 3) demonstrated the demand for and easy adoption of cashless person-to-person transactions. As early as 2009, a report cosponsored by the Pacific Financial Inclusion Programme and International Finance Corporation outlined the benefits (and challenges) of mobile money in PNG (Bruett and Firpo 2009). Mobile money potentially overcomes problems such as the safety risk of storing cash and the inconvenience of travelling long distances to town to visit banks or pay bills. In 2009, Digicel launched its popular Easipawa Easipay service, whereby subscribers could purchase electricity with their prepaid Digicel airtime credit. The service subsequently also became available through the bmobile network and commercial mobile banking accounts.

Subscribers could recharge their PNG Power meters at any time of day by using an SMS voucher sent to their phone instead of making long trips to crowded stores with limited business hours in order to purchase vouchers. The positive response to Easipawa Easipay—some 68,000 Digicel customers in 2009 (Oxford Business Group 2012c)—inspired optimism in the future of mobile money in PNG.

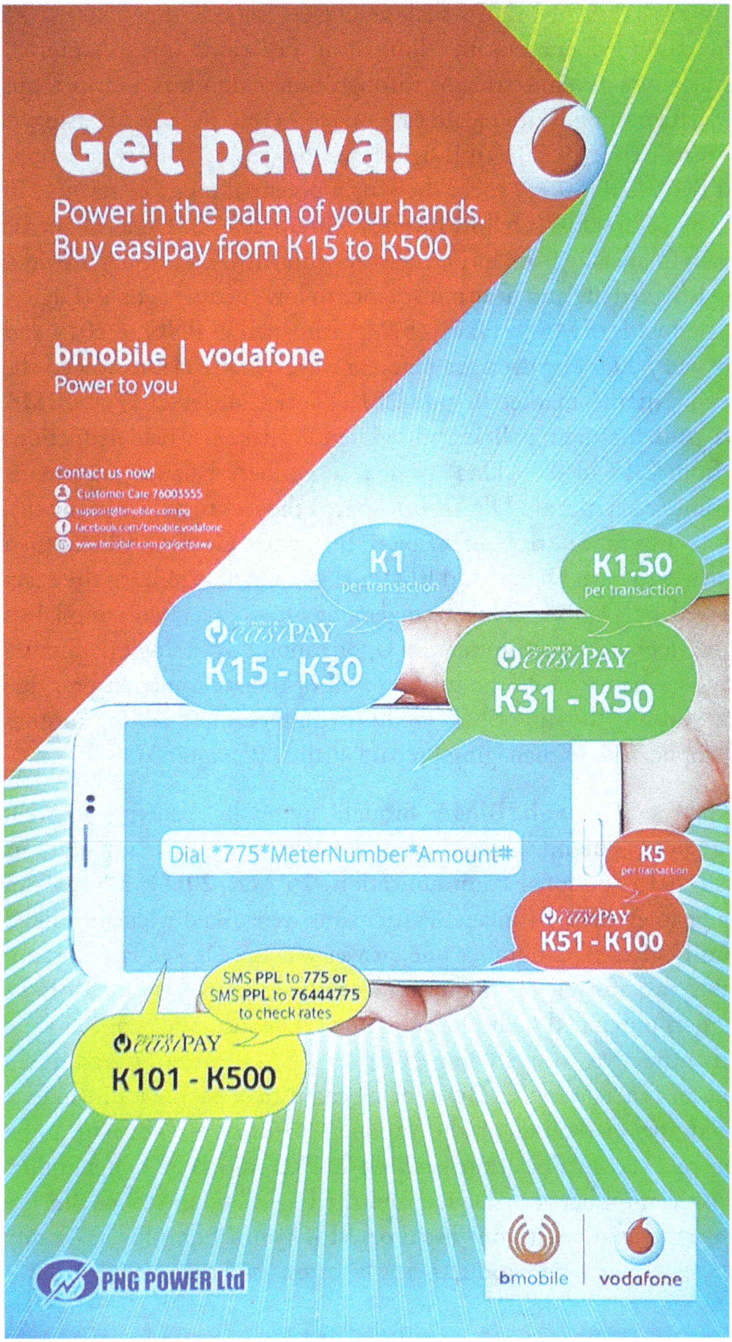

Figure 6.1. Ad for PNG Power's Easipay service at bmobile-Vodafone retail store, Port Moresby, 2018.
Source: Photo by R Foster.

From the perspectives of MNOs such as Digicel, mobile transactions can reduce the cost of operations by eliminating flex cards (see Chapter 3) and introduce new revenue streams through transaction fees and, perhaps eventually, data mining. With regard to transaction fees, the case of Easipawa Easipay is instructive: 'By 2010 it [Digicel] was processing 7000 transactions per day at PGK0.70 ($0.33) each, showing a potential monthly revenue of PGK147,089 ($70,000)' (Oxford Business Group 2012c). Digicel likewise stood to gain financially from its partnership with BIMA, a leading provider of mobile-delivered life and health insurance to low-income people (Digicel also partnered with BIMA in Haiti). BIMA launched in PNG in 2014 and reportedly had 300,000 customers before a wave of false claims forced the company to exit the market in mid-2019 (Post-Courier 2017c). BIMA customers paid for their policies with Digicel airtime; small deductions (ranging from 0.18 to 0.54 kina) from a subscriber's balance were made about 20 times each month. Digicel transferred the fees to a third company, Capital Life Insurance (part of Capital Insurance Group), which then remitted shares to both Digicel and BIMA. (I was not able to learn the exact share received by Digicel.) This arrangement often generated complaints when a subscriber had no airtime credit. Upon topping up, a single deduction from the new balance would be made for all of the accumulated past due deductions, causing the subscriber's credits to disappear and adding another layer of complication to managing prepaid airtime (Chapter 3).

Digicel's partnership with BIMA brought symbolic benefits as well, enhancing Digicel's brand.[2] According to the director of BIMA's operations in PNG in 2015 (personal communication, 25 May 2015), Digicel ran focus groups to evaluate its image. Participants were asked what they liked most about Digicel. The number one answer was the Digicel Foundation. The number two answer was the insurance policies offered through BIMA that customers purchased with Digicel airtime. Life insurance policies were particularly attractive given the high costs in PNG of meeting traditional obligations for funeral services. Both the mechanism of paying for insurance policies and the marketing of the products themselves encouraged the perception of a close association between Digicel and BIMA, not unlike the close association between Digicel and the Digicel Foundation. But the first of the terms and conditions listed on BIMA's brochures for its 'family life' and '*hausik* insurance' products clearly states: 'None of Digicel (PNG)

---

2   In 2018, MiBank, a leading innovator in micro finance and digital financial inclusion in PNG, also entered into partnership with BIMA.

Limited, its parent company, its subsidiaries or their related corporate bodies are liable for any benefits payable under this insurance.' Despite the value of the partnership for Digicel's reputation as a good corporate citizen, the company was legally careful to present itself as only the exclusive facilitator of premium payments to BIMA.

Digicel PNG, because of its dominance of the mobile market and presence in rural areas, provided the platform for all mobile money initiatives, such as the mobile banking services offered by the two main commercial banks, Bank South Pacific (BSP) and Kina Bank. (Kina Bank absorbed the operations of Australia and New Zealand Banking Group [ANZ] in 2019; the proposed sale of Westpac's operations to Kina was blocked by PNG's Independent Consumer and Competition Commission [ICCC] in 2021.) This arrangement worked to Digicel's financial advantage. For example, one manager involved with Westpac's instore banking program explained how customers could deposit and withdraw money, using a plastic card and a store's EFTPOS machine, without travelling to bank branches. Customers could keep track of their accounts on a basic mobile phone using simple text commands, but the bank paid a small fee to Digicel whenever a customer logged in to the USSD (unstructured supplementary service data) system. Each time a customer checked their balance at an ATM, a frequent practice for anxious customers awaiting cash transfers, Westpac paid Digicel a fee of 0.08 kina (personal communications, 2015, 2018). Accordingly, Westpac reduced the number of times per day that a customer could check their balance for free from six to three.

Digicel's arrangements with mobile money operators (MMOs) also advanced the company's long-term goal of reducing 'the cost of operations on the recharge side of the business by eliminating voucher cards and by passing [sic] air-time sellers' (Telepin Software Systems, Inc. n.d.). For example, an officer of MiBank reported in 2015 that in about four years of operation, the bank had processed about 846,000 transactions from the mobile wallets of its customers. Of these transactions, about 655,000 were airtime top-ups; that is, MiBank customers were using their mobile wallets to debit their accounts in order to purchase phone credit either for themselves or for others. The average top-ups were 3.80 kina for self and 3.25 kina for others.

Figure 6.2. Westpac instore banking at urban settlement near Port Moresby, 2016.

Source: Photo by R Foster.

Digicel also offered its own virtual wallet product, Cellmoni, prompting questions of conflict of interest—that is, whether MMOs partnering with Digicel can expect that their service will be prioritised (see Kachingwe and Berthaud 2014). A very brief review of Cellmoni's history suggests a few of the challenges involved in promoting mobile money in PNG, which in turn bring to light tensions in the moral economy of mobile phones. Cellmoni was introduced in PNG in 2011, including a rollout in Goroka, where there was an expectation that small coffee farmers would rapidly adopt the product. (A similar expectation motivated the choice of Goroka as one of the sites for the Australian Research Council research project on which this book is based.) Hoping to duplicate the well-known success of M-PESA in Kenya, Digicel anticipated that Cellmoni users would buy airtime credit, transfer funds to other Digicel users, pay bills and deposit and withdraw funds, visiting Digicel agents as needed. The service was promoted with animated television advertisements that highlighted the convenience of all these features.[3]

---

3   See: www.youtube.com/watch?v=Bh5T9IGN2QM, accessed 27 December 2022.

In March 2015, according to a Digicel executive, Cellmoni had about 95,000 subscribers, but only 8,500 were active (personal communication, 18 March 2015). It was clear that Cellmoni had not gained the traction that Digicel had expected, especially in rural areas. The most enthusiastic users of Cellmoni were 'large companies and organisations that need to make payments (of royalties, salaries, etc.) to large numbers of people' (APEC Policy Support Unit 2016: 8). What happened? The Digicel executive whom I interviewed offered a few explanations. First, many people who register for the service fail to use it subsequently because they do not know how. The issue here, it was claimed, was one of consumer education. Second, many people who registered—as many as 20 to 30 per cent—lost the numbers they needed to access their accounts. Other people were apparently reluctant to use Cellmoni because they feared losing their money if they lost their phone, an exceedingly common occurrence in PNG. Third, liquidity is a problem for many Cellmoni agents. That is, subscribers cannot always easily withdraw cash when desired because agents—for example, trade store owners—have limited amounts of cash on hand and require cash reserves for other business transactions. Transporting cash to restock agents who have run out is both costly and dangerous.

The problem of establishing a reliable and accessible network of agents for deposits and withdrawals has bedevilled almost all attempts to extend mobile money initiatives into rural areas of PNG. But the problem afflicts urban areas as well. I was able to witness different elements of this problem on two trips accompanying managers from different commercial banks in charge of mobile money projects, one based in Goroka and the other in Port Moresby. The first trip involved stops at two trade stores in East Goroka that acted as mobile money agents and where purchases could be made using a mobile access account. There had been no activity in recent months at either location. At the first and larger location, it was determined that the owner had gone abroad on holiday, taking with them the mobile phone used for the store's account rather than entrusting the phone to a worker. At the second location, it took some time to retrieve the phone used for the store's account, which was eventually found at the rear of the shop among other phones brought in by customers to have batteries recharged for a fee. When the bank manager attempted to test the phone by making a small purchase using the manager's own mobile access account, it became clear that the clerk had forgotten how to process the transaction. The manager patiently took the clerk through the various steps required. On the second trip, to a location outside Port Moresby that had not been visited lately, the bank manager discovered that the store's management had changed hands. The store was no longer providing instore banking services, and the expensive EFTPOS machine used to process transactions had disappeared along with the previous store manager.

In different ways, the experiences of these two trips remind us that there is an irreducible material substratum necessary for the operation of mobile money systems that proclaim the superiority of virtual or digital money to hard cash—paper currency and metal coins. Mobile phones and EFTPOS machines must be present, with charged batteries and in the hands of knowledgeable operators for the system to work. These experiences also remind us of the centrality of trust in the operation of mobile money systems (any money system, really; see Hart 1986; Foster 1999), including the trust put in agents, or not, to operate both honestly and expertly. One Digicel financial officer planning in 2016 for a relaunch of Cellmoni acknowledged that building trust was crucial for the success of the product (personal communication, 6 April 2016). He acknowledged that if a customer visits a Cellmoni agent and is told the system is down or there is no cash on hand, then trust is eroded. That is, the company has not met its moral obligation to its loyal consumers. According to a 2019 GSMA Intelligence report:

> As of August 2018, Digicel was processing more than 1 million transactions per month, primarily through its bank partners. However, the operator is looking to leverage its extensive distribution network to reach the last mile and expand rural financial inclusion. Digicel has plans to relaunch its mobile money service, leveraging its agents selling airtime top-ups as mobile money agents. (Highet et al. 2019: 27)

Digicel Cellmoni agents, it was hoped, would be able to mitigate the persistent problem of running out of cash by taking in cash through the sale of airtime.

This particular financial officer also worried about the expression of mistrust on the Digicel Complaints Group Facebook page (see Chapter 4) and the widespread perception that Digicel does not accurately calculate one's data balance. He noted the obsessive way in which some subscribers continually checked their balances (for which a tiny fee was charged by Digicel) and proposed that subscribers should not have to do this, even when waiting for a money transfer. Instead, the subscriber should receive an SMS (text message) confirming both the transfer and the amount transferred. Making fees more transparent, he claimed, would build trust and perforce strengthen the moral economy of mobile money services.

Digicel relaunched its mobile money (electronic wallet) service as CellMoni in January 2020. The opportunity to sign up for a CellMoni wallet was provided as part of the process for SIM card registration, perfectly illustrating how attaching unique identities to unique numbers (see Chapter 5) furthered the mission of 'financial inclusion'. Advertisements in newspapers and on Facebook encouraged consumers to use CellMoni to buy PNG Power Easipay and transfer money to other CellMoni users (peer to peer [P2P] transactions) without paying a fee as well as to buy Digicel airtime or a Digicel TV plan. A bonus of 25 per cent was offered for top-ups between 1 kina and 100 kina made with CellMoni. (A January 2020 newspaper report suggested that cash withdrawals were not yet available but that this important feature would be added later [Post-Courier 2020].)

In October 2020, Digicel partnered with Coca-Cola Amatil Ltd in a familiar style of promotion that gave consumers a chance to win between 2 kina and 1,000 kina of CellMoni. Consumers were called upon to purchase a 500 mL plastic bottle of Coca-Cola, peel back the label, find a 12-digit number, and text the number to 444 in order to determine if it is a winner. (It is unclear from the promotional advertisement if winners were able to cash out their prizes.)

> Digicel PNG chief executive Colin Stone said the promotion, through its product CellMoni, was a continuation of the work they do.
>
> 'For us, CellMoni is a continuation of what Digicel started when we launched here in Papua New Guinea in 2007, and that's around connecting the unconnected,' he said.
>
> 'CellMoni is a way of helping the unbanked (population) become banked, helping them access financial services and improve financial literacy across the country'. (National 2020e)

Stone's comments suggest no reservations about the 'unconnected' becoming 'connected' by means of the purchase of a non-nutritious drink of sugary carbonated water. His comments, moreover, betray no awareness, ironic or otherwise, of the way in which a foreign-owned telecommunications company was representing itself as an agent of national development, extending to the people financial services that the state on its own was incapable of providing. In PNG, the state—in the form of the Central Bank—instead creates a regulatory environment in which telecommunications companies, partnered with banks and supported by grants and aid from the Asian

Development Bank, European Union, World Bank group and so forth, do 'social good' convincing people to improve their lives by entering the formal financial sector (see Taylor and Horst 2018).

***

The discourse of financial inclusion is, like all developmentalist discourse, a moral one. Entering the formal financial sector is the first step toward enhanced wellbeing, not only for particular individuals but also for the nation as a whole. By signing up for banking services, the horde is rehabilitated by making its tiny hoards available as savings in the form of capital holdings (Peebles 2014). Refusal or inability to take that first step is understood as a deficiency, whether technical or ethical. In any case, the remedy is always better pedagogy—more effective programs of financial literacy and consumer education. The failure of mobile money initiatives in PNG, including the initial launch of Digicel's Cellmoni in 2011, is thus commonly attributed to lack of consumer awareness.

While there is undoubtedly truth in the claim that many consumers need better instruction and support in their use of mobile money, there are also other constraints on the success of mobile money initiatives. For example, many of these initiatives—including the microinsurance program run by BIMA and Digicel in PNG—have been underwritten with the resources of development agencies and NGOs, most notably the Pacific Financial Inclusion Programme launched in 2008 and jointly administered by the United Nations Capital Development Fund and the United Nations Development Programme. One commercial bank manager pointed out that his bank's interest in projects such as instore banking waxes and wanes with the availability of external funding. The bank is conservative, he claimed, willing to take risks with other people's money but not its own (personal communication, 26 June 2018).

Other limitations suggest that the challenge of launching mobile money networks in rural and peri-urban areas is not so much a matter of changing mindsets as it is establishing a stable network of agents for making deposits and withdrawals. One microfinance bank manager who several years earlier had identified consumer education as a real problem claimed in 2018 (personal communication, 25 June 2018) that one can talk about digital financial literacy all one wants, but getting an agent network in place is the key to making DFS work. This claim echoes the opinion expressed in 2013 by Tony Westaway, then a manager at Nationwide Microbank (subsequently CEO of MiBank):

I think the regulator will need to step in and promote inter-operability between the wallets in order that everyone can participate and all Papua New Guineans whether they have got a Digicel phone or a Bmobile phone or a BSP WantokMoney or Nationwide Microbank Micash that they can send money to each other and share agents. (Islands Business 2013)

In other words, the state still has an important role to play in creating the regulatory environment—which in PNG is regarded as favourable for innovation (Telepin Software Systems Inc. n.d.)—for the expansion of mobile money. Both telecoms and banks, whose interests do not always align (see Mas and Staley 2014), thus stand on the moral grounds of greater financial inclusion in order to place claims on the state.

The same bank manager who acknowledged the prime importance of a stable agent network also candidly wondered about the premises of not only teaching poor people digital financial literacy but also aiming DFS at poor people. He confided that a well-respected politician once asked him whether poor people ought to instead be taught how to create wealth: They have no money—what is the point of learning how to budget? Perhaps, then, the providers of DFS were targeting the wrong group at the bottom of the pyramid. He wondered if these efforts ought to be aimed instead at people with some resources and assets, people who have something financial to be included in the first place.

The Moral Economy of Natural Disaster

The Pacific Islands region has since 2011 experienced several devastating natural disasters, including droughts, cyclones, tsunami and earthquakes. Climate change promises more frequent occurrences of extreme weather events. These precarious circumstances have offered further justification for expanding mobile money initiatives. For example, in Vanuatu, 'the most at-risk country on earth to natural disasters' (McGarry 2020), Oxfam has pioneered a digital project in order to shift the mode of delivering humanitarian assistance from goods to cash. The project, called UnBlocked Cash, established a decentralised cashless payment system, powered by blockchain technology. Money loaded on to plastic cards can be used by registered participants for purchases at local stores by tapping on the Android mobile phones given to registered vendors. Participants do not need a bank account, although vendors do, and are free to spend their funds as they see fit. This approach aligns well with Prahalad's (2010) vision of improving the quality of life for people at the bottom of the pyramid by empowering them as consumers. According to Oxfam's head of Pacific cash and livelihoods in Vanuatu, the approach 'is more market-friendly, [and] gives people more dignity and flexibility' (McGarry 2020).

Disasters can also afford opportunities to advance corporate agendas (see Klein 2007). The COVID-19 disaster that hit the Pacific Islands region in 2020 offered an opportunity for Digicel both to highlight its role in emergency response and to link this role to the regulatory terms of the company's relationship with the state. In September 2020, Digicel Group Chairman Denis O'Brien authored a case study issued by the Broadband Commission for Sustainable Development, an organisation:

> established in 2010 by ITU and UNESCO with the aim of boosting the importance of broadband on the international policy agenda, and expanding broadband access in every country as key to accelerating progress towards national and international development targets. (Genéve Internationale n.d.)

The case study noted Digicel's philanthropic efforts, made through the Digicel Foundation, to distribute personal protective equipment and sanitary equipment. It also noted the efforts undertaken as an employer to introduce safety measures at work and to facilitate working from home. But the case study mainly focuses on the collaboration of the company with government agencies in responding to COVID-19, and on the company's assistance in disseminating messages and expanding connectivity. This assistance, O'Brien argued, makes the case for a particular vision of how development ought to unfold in places like PNG.

With respect to communications, O'Brien (2020) observed: 'A number of Governments sought our support in ensuring that messages could be quickly disseminated to citizens and that *reliable* information from *trusted* sources could be readily accessed.' In PNG, Digicel set up a toll-free number to broadcast COVID-19 messages from the Department of Health. It also changed the network identifier displayed on mobile devices to '*wasim han gut Digicel*' ('wash your hands good Digicel') to reinforce the department's messages about practising basic hygiene. In other markets, according to O'Brien, the company repurposed its SMS and online marketing platform to communicate information about practising basic hygiene. It is unclear whether this service was provided free of charge, as Digicel PNG has previously been accused of declining to transmit tsunami warning messages unless compensated at commercial rates (Chamberlain 2019; see Watson 2012).

O'Brien (2020) drew a pointed lesson from the fact that 'Governments want to use traditional telecoms channels to communicate important information': that these channels are more trustworthy than information that circulates though online social media. In other words, telecoms channels offered an antidote to 'fake news' and misinformation such as the anti-vaccination rhetoric that coincided with an outbreak of measles in Samoa in 2019. Although reports of misinformation related to COVID-19 on social media have appeared—some reports linking symptoms of COVID-19 to 5G (see Horst and Foster 2024)—at least one study by anthropologists suggests how a Facebook group 'played an important and positive role in the ways in which many people engaged with what could have emerged as a health disaster' (Dwyer and Minegal 2020: 233). O'Brien's larger concern, however, was less with COVID misinformation and more with the platforms on which such misinformation circulated, a concern that again raises the question of who pays for telecommunications infrastructure and thus the means of connecting the unconnected in developing countries like PNG (Chapter 4).

O'Brien's invidious comparison between online fake news and trustworthy text messages reflects his view that not all communication is of equal value. COVID-related lockdowns that entailed working and schooling remotely increased the demand for capacity on mobile networks. In order to meet this demand, O'Brien noted, governments not only made additional spectrum available but also waived the associated fees. MNOs, in turn, 'put in place traffic management measures to protect limited capacity from the upswing in "entertainment" content' (O'Brien 2020). In some instances, Digicel 'worked with Governments to zero-rate the data associated with on-line education and Government information websites', thus expanding online connectivity to people otherwise excluded due to economic hardship. These initiatives, O'Brien (2020) concluded, demonstrated that while some data directly contributes to the wellbeing of a country and its citizens, other data is of 'dubious provenance that enriches only overseas platforms'.

The response by Digicel and governments to COVID-19 thus bolstered O'Brien's arguments against net neutrality and his protests against so-called over-the-top (OTT) players such as Google, Facebook and Skype (Chapter 4). O'Brien has asserted that 'net neutrality is great for those living in the United States but not for those living in Africa' because OTT players can use a developing country's networks without making any financial contribution. Whereas, according to O'Brien, MNOs invest 25 per cent of their revenues in capital expenditure on infrastructure, 'OTT players and content providers generally have a free ride' (ITU News 2015: 16). Governments, moreover, need to support MNOs by abandoning the practice of selling spectrum and new licences 'for vast amounts of money' (ITU News 2015: 17). In O'Brien's view, governments and regulators 'see telecom operators as profitable businesses' and organise spectrum auctions accordingly. But in developing countries, spectrum fees take money out of the industry. Instead of levying charges, O'Brien would prefer:

> contractual obligations on operators to roll out the networks quickly, and if operators fail to deliver. Then by all means, penalize the operators for not fulfilling the contractual obligations they signed up to. (ITU News 2015: 17)

* * *

In sum, COVID-19 allowed Digicel not only to articulate but also to implement O'Brien's vision of post-political governance: an ever-expanding broadband ecosystem based on public–private partnerships and 'philanthrocapitalism' of the sort represented by the Digicel Foundation. 'Mobile money solutions' are very much part of that vision, and their value was particularly salient during lockdowns in which people could make payments and purchase electricity remotely. Government must play its part here, too, if connecting the unconnected is to happen: 'Regulatory obstacles to the introduction of mobile money solutions have the effect of excluding the most disadvantaged persons, who are without bank accounts or credit cards' (O'Brien 2020). Companies grow their business and profits, states improve the lives of citizens, and consumers enjoy the multiple benefits of connectivity—convenience, security and savings. O'Brien's vision is nothing less than a twenty-first-century version of the moral economy conjured up by Adam Smith's enduring image of the invisible hand.

Conclusion: Infrastructural Citizenship and Uneven Connections

Paula Uimonen (2015: 37)has observed with regard to Africa that:

> In the absence of other forms of modern infrastructure (e.g. educational institutes, health care facilities, roads and transport systems), mobile infrastructure offers at least a semblance of accomplishment, offering citizens a sense of progress and inclusion. (See also Fredericks and Diouf 2014; Wafer 2012)

The arrival of Digicel in Papua New Guinea (PNG) indeed offered many citizens 'a sense of progress and inclusion'. Put differently, Digicel seemed poised to realise the dream of 'coevalness' (Fabian 1983) implicit in Benedict Anderson's (1991) well-known formulation of the nation as an imagined community: one day, all citizens would be connected in an experience of national simultaneity. The PNG state had apparently at last delivered the desirable kind of modern and inclusive 'infrastructural citizenship' (Lemanski 2018) that formal Independence in 1975 had anticipated—albeit by relinquishing the state's monopoly on telecommunications and creating a competitive market.[1] Such infrastructural citizenship would accordingly materialise a healthy moral economy in which the PNG state (via the state-licensed company Digicel) and its citizens (and consumers) all lived up to their obligations to each other.

---

1 As Lemanski (2022: 936) puts it, 'the state is materially and visibly represented through everyday (in)access to public infrastructure, while the state imagines and plans for citizens primarily as infrastructure claimants, consumers and complainers'.

The years immediately following Digicel's arrival were a period of increasing digital equality and expanding infrastructural citizenship. Inclusion in the national space-time was extended to more people than ever before, and the possibility of a future in which just about everyone was included seemed thinkable. It was during this period that Digicel enjoyed almost universal affection among Papua New Guineans, at least by comparison with the deep popular disdain for Telikom PNG. Fast forward: 10 years later in 2017, the national unity that Digicel had gone some way toward realising was giving way to new forms of digital division. The promise of coevalness was undermined by infrastructural developments that left all people in some places and some people in all places 'out of sync' (Pype 2021), involuntarily disconnected from the synchronicity promised by digital technology in general and by Digicel in particular. Whereas for a brief moment it seemed like a one-to-one correspondence might emerge between physical space and Hertzian space, the two topographies were becoming disjunct from each other (see Meese et al. 2019).

In other words, infrastructural citizenship with its attendant claim to national belonging was becoming more differentiated in form and degree, entrenching an older digital divide between urban and rural areas (see Curry et al. 2016) that deferred if not denied the dream of national coevalness (Fabian 1983). Infrastructural citizenship was experienced as a function of one's location inside or outside of a network of uneven connections. That is, not all citizens enjoyed equal connectivity. For people living in PNG's so-called last places (*las ples* in Tok Pisin), this condition entails a kind of perpetual 'suspension', in which the 'ever-present gap between the start and completion of infrastructure projects' is never closed (Gupta 2015). Instead, people occupy 'the suspension between what was promised and what will actually be delivered' (Gupta 2018: 70).

***

Fifteen years after Digicel launched, the infrastructural assemblage that makes mobile communication possible for PNG's citizens faced changes to its composition and challenges to its stability. First, in July 2022, Australia's geopolitical infrastructural intervention (Chapter 1) continued when a subsidiary of Telstra Corporation Ltd completed its acquisition of Digicel's operations in the Pacific Islands (Digicel Pacific Ltd), including Digicel's largest Pacific market in PNG. Telstra, an Australian telecommunications company, contributed USD270 million to the deal, while the Australian Government contributed USD1.33 billion.

The *Financial Review*, like many other commentators, noted that the Australian taxpayers were effectively assuming the risks involved in the purchase, but observed:

> this is because of fears that Digicel, which came under immense pressure when mobile phone traffic plunged in the tourist-dependent Pacific region at the height of the pandemic, may have been used to spy on Australia's neighbours if it fell into Beijing's hands. (Baird and Tillett 2022)

Other commentators similarly pointed out that Telstra benefitted handsomely from the Australian Government's huge subsidy, but also wondered whether the deal would prove beneficial for the people of PNG (Howes 2021; Sora and Pryke 2021). Would Telstra operate its mobile network in PNG in such a way as to include more of the population by reducing prices and expanding access to rural areas? Would Telstra, in other words, extend full infrastructural citizenship to people excluded by changes in the telecommunications network—the shift from telephony to data— or to people never included to begin with, and thus restore the promise of national connectivity?

Second, a hopeful future of greater connectivity was resuscitated by the debut in April 2022 of a credible potential competitor to Digicel in PNG: Digitec Communications Ltd, trading as Vodafone PNG.[2] CEO Pradeep Lal spoke the corporate language of projected growth:

> According to GSMA reports, market penetration here is about 37 per cent of the population … Fiji is at 130 per cent mobile penetration—the same as Australia and New Zealand. Vanuatu and Samoa are at 100 per cent. PNG can very easily reach between 80 and 90 per cent mobile penetration in the next couple of years. (Business Advantage PNG 2022)

From this point of view, Pradeep could interpret the absence of his product in the market as an opportunity rather than a liability. Making that product available will require the construction by Vodafone PNG of its own network of towers, given the lack of provisions for tower sharing in PNG.

---

2   Vodafone PNG (Digitec Communications Limited) is a subsidiary of Amalgamated Telecom Holdings (ATH) and Austel Investment Pty Limited. 'ATH has delegated the operation and management of the business to its subsidiary Vodafone Fiji Pte Limited, the largest mobile telecommunications provider in Fiji' (vodafone.com.pg/about, accessed 14 September 2022).

Vodafone PNG's promotion of its rollout in 2022 recalled Digicel's 2007 strategy of tracking its step-by-step expansion across the national territory in newspaper advertisements. Vodafone PNG documented its premiere in Port Moresby and then its roadshow teams working in other parts of the country, although Vodafone PNG's promotion unfolded on Facebook rather than in printed media. The same social media platform, however, afforded users the opportunity to question the reality of a slowly growing area of even coverage implied in Vodafone PNG's promotions. A message announcing a new tower and welcoming Vodafone's service to villages in the Aroma Coast area south of the national capital was recirculated with a complaint about blackspots in Vodafone's coverage of Port Moresby. Some commenters complained about slow internet speeds or lack of reception in particular locations, and one commenter joked that the Aroma villagers only need the tower for voice calls not data calls, thus deeming obsolete technology appropriate for rural areas (see Foster 2023). These complaints highlight the experience of uneven connections and suggest that infrastructure is indeed an incomplete process, but not quite in the same way that Pradeep Lal, thinking of penetration rates, might have in mind.

***

The question of infrastructural citizenship bears consideration in light of the 2016 United Nations declaration that access to the internet is a human right (Howell and West 2016). This declaration focused on governmental responsibility for ensuring that citizens are not denied access to the internet as a means for free political expression; but it did not obligate governments to provide access to all citizens, especially the poorest. Despite efforts like that of the Mexican Government to make good on its 2013 constitutional amendment defining access to the internet as a human right (Barry 2020), approximately one third of the world's population in 2022 did not use the internet (ITU n.d.).

In PNG, according to the report of Highet et al. (2019), access to the internet and more generally to basic telecommunications services was highly uneven in 2018. In a population conservatively estimated by the report at 7.6 million, there were 2,525,643 unique subscribers—a market penetration rate of almost 30 per cent. Almost 1 million mobile internet unique subscribers yield a market penetration rate of 11.75 per cent. But the majority of connections by mobile technology were 2G (55.54 per cent), with 3G connections at 22.91 per cent and 4G connections at 21.55 per cent. In 2022, 2G connections had declined to 26 per cent, while both 3G and 4G connections increased to 31 per cent and 43 per cent, respectively, although the subscriber penetration rate of 41 per cent suggested that the majority of the population remained unconnected (GSMA 2023). These statistics, to repeat, reflect digital divides between urban and rural areas and between data users and voice users; in 2018, 70 per cent of internet users resided in the two major cities of Port Moresby and Lae (Highet et al. 2019: 24). It is primarily in urban areas, moreover, where mobile users enjoy bmobile/Telikom and Vodafone's relatively cheaper data plans (Chapter 1, footnote 18). The situation in both 2018 and 2022 was thus one of highly varying forms and degrees of connectivity across PNG, especially broadband connectivity.

As Logan and Forsyth (2018: 18) note, access to the internet—largely via mobile phones—is crucial in countries such as PNG for development-related projects 'in diverse areas such as maternal health, microfinance, and teacher education'. Availability and affordability of mobile phones are thus increasingly prerequisites of full infrastructural citizenship. In PNG, however, it is the corporation and not the state whose actions bear upon the lived reality of infrastructural citizenship when it comes to telecommunications:

> In a country like PNG where the state does not have the capacity or the will to address infrastructure deficiencies, corporate interests have been able to insert themselves into citizens' lives, for good or bad, introducing great change but also precariousness. (Logan and Forsyth 2018: 20)

Digicel PNG (now Telstra), which controlled upwards of 90 per cent of the mobile market and owned its towers and terminals, rather than the PNG state, determined the terms and conditions on which infrastructural citizenship and the promise of national coevalness and being in sync were offered. And despite its nontrivial investment in corporate social responsibility through the activities of the Digicel Foundation in building schoolrooms and promoting gender equity (Chapter 6), the company always remained a for-profit business. It was guided in the last instance by its commercial interests, which of course prompted the sale of Digicel Pacific Ltd to Telstra.

The acquisition of Digicel Pacific by Telstra reset the infrastructural clock. Telstra International CEO Oliver Camplin-Warner affirmed his company's commitment to 'building a strong and sustainable PNG' by announcing that an additional new 115 towers would be constructed across the country in the next two years (Post-Courier 2022b). He said:

> This investment will mean continued improvements to 4G coverage, particularly in rural areas, which will bring with it opportunities to improve health, education, agricultural, commerce and cultural outcomes through the use of technology.

Colin Stone, CEO of Digicel PNG, noted that 'Telstra has experience connecting regional and remote customers in challenging geographies across mountains, deserts, rainforests and coastlines' (Post-Courier 2022b). The dream of national coevalness was thus revived and its realisation projected into a near future.

***

The comments of Camplin-Warner and Stone also suggest something about the infrastructure for mobile telecommunications, specifically, in PNG and elsewhere—namely, the frequency with which upgrading is required. Maintenance of trunk infrastructure is always necessary, of course, even if less visible than initial construction (Gupta 2018). But once sewer or electrical lines are laid, they usually do not require constant upgrading that threatens to exclude some users from access to full service. This threat looms not only in PNG, but also in countries of the Global North such as the US, where the imminent shutdown of 3G wireless networks is anticipated to affect older and low-income Americans disproportionately (Zakrzewski 2021).

Constant upgrading puts infrastructural citizenship at risk. Moreover, telecommunications infrastructure is unlike other kinds of infrastructure inasmuch as it is difficult to hack. That is, it is one thing to jerry-rig connections to water pipes or electrical lines (see e.g. von Schnitzler 2008), or to build makeshift housing or squat on state-owned land—all creative forms of popular protest against limitations on infrastructural citizenship. It is another thing to build one's own telecommunications network or even to erect a cell tower: not impossible (see González 2020), but usually beyond the means of even the most insurgent of citizens. The infrastructural precariousness that Logan and Forsyth (2018) duly note is therefore not only a function of corporate control, but also a consequence of the material specificities of mobile telecommunications networks.

How can this precariousness be alleviated? Answers to this question highlight the formidable challenges that both ordinary Papua New Guineans and the PNG state face in negotiating the moral economy of mobile phones. On the one hand, Logan and Forsyth (2018) call for a reduction in the influence of corporations like Digicel in determining the shape of telecommunications networks. Digicel has over the years of its operation in PNG benefitted from low-cost loans offered by the World Bank's International Finance Corporation (IFC) (Chapter 1). But this type of loan is not available to governments, only to private sector–led projects. The latest state-led projects in PNG to upgrade the undersea fibre optic network have been made possible by dubious loans from China or—like the sale of Digicel Pacific Ltd to Telstra—by the largesse of Australian taxpayers. What remains to be seen is whether the price of such financial aid for extending infrastructural citizenship has been to render the PNG state and the people of PNG pawns in a game of geopolitical chess.

Logan and Forsyth wish for more state regulation rather than less, but worry that 'Papua New Guineans are at the mercy of a corporation, over which their own government appears to have limited control' (Logan and Forsyth 2018: 20; see Chapter 4). On the other hand, billionaire Denis O'Brien, Digicel Group Ltd's founder and chairman, has called on governments to lower the barriers to private investment, such as high spectrum licence fees (ITU News 2015). In his role as a member of the UN Broadband Commission for Sustainable Development, moreover, O'Brien strenuously championed the idea that access to the internet is a human right. But O'Brien represented Digicel as a victim of other corporations in realising this goal. He demanded that over-the-top (OTT) players such as Google and Facebook (now Meta), who free ride on the rails laid down by mobile network operators, pay their fair share (see Chapter 4). It is difficult to decide whether Logan and Forsyth's vision of effective regulation by the PNG state or O'Brien's proposal for Google and Facebook to subsidise telecom companies is less likely to be realised.

References

ABC. 2020. 'PNG's Plans to Deactivate all Unregistered Mobile Phones "Must Proceed," Communications Official'. *Pacific Beat*, 20 January. www.abc.net.au/ pacific/programs/pacificbeat/png-plans-to-deactivate-all-unregistered-mobile-phones/11884872, accessed 6 December 2022.

ABC International Development. 2019. *PNG Citizen Perceptions of Governance and Media Engagement Report*. Port Moresby, Papua New Guinea: ABC International Development. live-production.wcms.abc-cdn.net.au/54df1da6a 4dbe1bdb212ec10b655ec4c, accessed 31 May 2023.

Abe, Thomas. 2007. 'Time to get the Facts Straight on Mobile Competition'. *National*, 31 July.

ADB (Asian Development Bank). 2012. 'New Telecommunications Network in Papua New Guinea'. www.adb.org/results/new-telecommunications-network-papua-new-guinea, accessed 28 November 2022.

ADB. 2019. 'ADB's First Satellite Financing to Expand Internet Access in Asia and Pacific'. News release, 2 December.

ADB. 2020. *Bemobile Limited, Bemobile Expansion Project (Papua New Guinea and Solomon Islands) (Regional)*. Extended Annual Review Report. Asian Development Bank. www.adb.org/sites/default/files/project-documents/44937/44937-014-xarr -en.pdf, accessed 20 December 2022.

Alphonse, Stanley for Thomas Abe. 2008. 'The Role of Competition in Promoting ICT Development and the Responsibilities of ICT as Competition Regulator'. In *Information and Communication Technology (ICT): 'Into the Future for PNG!' Proceedings of All Day Workshop held at UPNG on 23rd January 2008*, 107–11. Port Moresby: Institute of National Affairs.

Anand, Nikhil, Akhil Gupta and Hannah Appel (eds). 2018. *The Promise of Infrastructure*. Durham: Duke University Press. doi.org/10.1215/978147800 2031

Andersen, Barbara. 2013. 'Tricks, Lies and Mobile Phones: "Phone Friend" Stories in Papua New Guinea'. *Culture, Theory and Critique* 54 (3): 318–34. doi.org/10.1080/14735784.2013.811886

Anderson, Benedict. 1991. *Imagined Communities: Reflections on the Origin and Spread of Nationalism.* Rev. ed. New York: Verso.

APEC Policy Support Unit. 2016. *Case Study on the Role of Services Trade in Global Value Chains: Telecommunications in Papua New Guinea.* September 2016. Singapore: APEC Policy Support Unit. www.apec.org/docs/default-source/publications/2016/10/case-study-on-the-role-of-services-trade-in-global-value-chains-telecommunications-in-papua-new-guin/216_psu_png-telcoms-and-gvcs-case-study_final.pdf?sfvrsn=7da4a006_1, accessed 25 July 2023.

Appadurai, Arjun. 1986. 'Theory in Anthropology: Center and Periphery'. *Comparative Studies in Society and History* 28 (2): 356–61. doi.org/10.1017/S0010417500013906

Appel, Hannah, Nikhil Anand and Akhil Gupta. 2015. 'Introduction: The Infrastructure Toolbox'. Theorizing the Contemporary. *Society for Cultural Anthropology,* 24 September. culanth.org/fieldsights/introduction-the-infrastructure-toolbox, accessed 8 July 2019.

Appel, Hannah, Nikhil Anand and Akhil Gupta. 2018. 'Temporality, Politics, and the Promise of Infrastructure'. In *The Promise of Infrastructure,* edited by Nikhil Anand, Akhil Gupta and Hannah Appel, 1–38. Durham: Duke University Press. doi.org/10.2307/j.ctv12101q3.4

Archambault, Julie Soleil. 2011. 'Breaking up "Because of the Phone" and the Transformative Potential of Information in Southern Mozambique'. *New Media & Society* 13 (3): 444–56. doi.org/10.1177/1461444810393906

Archambault, Julie Soleil. 2013. 'Cruising through Uncertainty: Cell Phones and the Politics of Display and Disguise in Inhambane, Mozambique'. *American Ethnologist* 40 (1): 88–101. doi.org/10.1111/amet.12007

Archambault, Julie Soleil. 2017. *Mobile Secrets: Youth, Intimacy and the Politics of Pretense in Mozambique.* Chicago: University of Chicago Press. doi.org/10.7208/chicago/9780226447605.001.0001

Bai Magea, Wendy. 2019. 'Making a Living in Urban Papua New Guinea: Community, Creativity and the Provision of Mobile Phone Goods and Services in Goroka'. BA(Hons) thesis, University of Goroka.

Bainton, Nicholas A. 2010. *The Lihir Destiny: Cultural Responses to Mining in Melanesia.* Canberra: ANU E Press. doi.org/10.22459/LD.10.2010

Bainton, Nicholas A. 2011. 'Are You Viable? Personal Avarice, Collective Antagonism and Grassroots Development in Melanesia'. In *Managing Modernity in the Western Pacific*, edited by Mary Patterson and Martha Macintyre, 231–59. St Lucia: University of Queensland Press.

Baird, Lucas and Andrew Tillett. 2022. 'Telstra Completes 2.4b Digicel Deal'. *Australian Financial Review*, 14 July. www.afr.com/companies/telecommunications/telstra-completes-2-4b-digicel-deal-20220714-p5b1h0, accessed 10 November 2022.

Baird, Timothy D. 2021. '"Wrong Number? Let's Chat" Maasai Herders in East Africa Use Misdials to Make Connections'. *The Conversation*, 24 June. theconversation.com/wrong-number-lets-chat-maasai-herders-in-east-africa-use-misdials-to-make-connections-156167, accessed 12 December 2022.

Balbi, Gabriele. 2017. 'Deconstructing "Media Convergence": A Cultural History of the Buzzword, 1980s–2010s'. In *Media Convergence and Deconvergence*, edited by Sergio Sparviero, Corinna Peil and Gabriele Balbi, 31–51. New York: Palgrave. doi.org/10.1007/978-3-319-51289-1_2

Bar, François, Matthew S Weber and Francis Pisani. 2016. 'Mobile Technology Appropriation in a Distant Mirror: Baroquization, Creolization, and Cannibalism'. *New Media & Society* 18 (4): 617–36. doi.org/10.1177/1461444816629474

Barker, Paul. n.d. 'Reform in PNG – The Digicel Story'. Unpublished manuscript. Port Moresby: Institute of National Affairs.

Barnett-Naghshineh, Olivia. 2019. 'Shame and Care: Masculinities in the Goroka Marketplace'. *Oceania* 89 (2): 220–36. doi.org/10.1002/ocea.5219

Barry, Jack. 2020. 'COVID-19 Exposes Why Access to the Internet is a Human Right'. *Open Global Rights*, 26 May. www.openglobalrights.org/covid-19-exposes-why-access-to-internet-is-human-right/, accessed 13 December 2022.

Barton, James. 2015. 'O3b Constellation Helps Digicel Meet Increasing Demand in Papua New Guinea'. *Developing Telecoms*, 16 June. developingtelecoms.com/telecom-technology/satellite-communications-networks/5885-o3b-constellation-helps-digicel-meet-increasing-demand-in-papua-new-guinea.html, accessed 28 November 2022.

Benson, Peter and Stuart Kirsch. 2010. 'Capitalism and the Politics of Resignation'. *Current Anthropology* 51 (4): 459–86. doi.org/10.1086/653091

Bower, Joseph L and Clayton M Christensen. 1995. 'Disruptive Technologies: Catching the Wave'. *Harvard Business Review* 73 (1): 43.

Brennan, Joe. 2019. 'Denis O'Brien Faces Pressure to Inject Equity into Digicel'. *Irish Times*, 29 November. www.irishtimes.com/business/technology/denis-o-brien-faces-pressure-to-inject-equity-into-digicel-1.4098467, accessed 25 July 2023.

Brennan, Joe. 2022. 'Digicel Completes a $1.6 Billion Pacific Unit Sale'. *Irish Times*, 14 July. www.irishtimes.com/business/2022/07/14/digicel-completes-16-billion-pacific-unit-sale/, accessed 25 July 2023.

Brimacombe, Tait, Romitesh Kant, Glenn Finau, Jope Tarai and Jason Titifanue. 2018. 'A New Frontier in Digital Activism: An Exploration of Digital Feminism in Fiji'. *Asia & the Pacific Policy Studies* 5 (3): 508–21. doi.org/10.1002/app5.253

Bruett, Tillman and Janine Firpo. 2009. *Building a Mobile Money Distribution Network in Papua New Guinea*. IFC Mobile Money Toolkit. Washington DC: World Bank Group. documents.worldbank.org/curated/en/1024715011393 55363/Building-a-mobile-money-distribution-network-in-Papua-New-Guinea, accessed 13 December 2022.

Bruns, Axel. 2008. *Blogs, Wikipedia, Second Life, and Beyond: From Production to Produsage*. New York: Peter Lang.

Bryden, Natalie. 2021. *Papua New Guinea: Telecoms, Mobile and Broadband – Statistics and Analyses*. Sydney: Paul Budde Communication Pty Ltd.

Business Advantage PNG. 2021. 'What are Telstra's Plans for Digicel Pacific?' 26 October. www.businessadvantagepng.com/what-are-telstras-plans-for-digicel-pacific/, accessed 1 December 2022.

Business Advantage PNG. 2022. 'Vodafone's K3 Billion Investment to Expand Papua New Guinea's Telco Market'. 27 April. www.businessadvantagepng.com/vodafones-k3-billion-investment-to-expand-papua-new-guineas-telco-market/, accessed 10 November 2022.

Business Wire. 2018. 'Digicel and Papua New Guinea Sustainable Development Programme Bridging the Communications Gap in Western Province'. 12 September. www.businesswire.com/news/home/20180912005681/en/Digicel-and-Papua-New-Guinea-Sustainable-Development-Programme-Bridging-the-Communications-Gap-in-the-Western-Province, accessed 7 December 2022.

Busse, Mark and Timothy LM Sharp. 2019. 'Marketplaces and Morality in Papua New Guinea: Place, Personhood and Exchange'. *Oceania* 89 (2): 126–53. doi.org/10.1002/ocea.5218

Callick, Rowan. 2015. 'State Takes Control of PNG Airwaves with Deal to Buy EM TV'. *The Australian*, 10 February.

Callon, Michel. 2021. *Markets in the Making: Rethinking Competition, Goods, and Innovation*. Translated by Olivia Custer. Brooklyn: Zone Books. doi.org/10.2307/j.ctv1mjqvf7

Callon, Michel, Cécile Méadel and Vololona Rabeharisoa. 2002. 'The Economy of Qualities'. *Economy and Society* 31 (2): 194–217. doi.org/10.1080/03085140220123126

Capey, Dev. 2013. 'Blogging for Social Change in Papua New Guinea: A Case Study Investigation of the Namorong Report'. Master's thesis, Auckland University of Technology.

Cave, Danielle. 2012. *Digital Islands: How the Pacific's ICT Revolution is Transforming the Region*. Sydney: Lowy Institute for International Policy.

Chai, Paul. 2020. 'Ringing in the Changes: SIM Registration the Biggest Challenge to bmobile's Ecommerce Future in Papua New Guinea'. *Business Advantage PNG*, 28 September. www.businessadvantagepng.com/ringing-in-the-changes-sim-registration-the-biggest-challenge-to-bmobiles-ecommerce-future-in-papua-new-guinea/, accessed 25 July 2023.

Chamberlain, Peter. 2019. *Independent Review of the 'Strengthening Disaster Risk Management in Papua New Guinea' Project*. www.dfat.gov.au/sites/default/files/sdrmp-review-report.pdf, accessed 25 July 2023.

Clark, Jeffrey. 1997. 'Imagining the State, or Tribalism and the Arts of Memory in the Highlands of Papua New Guinea'. In *Narratives of Nation in the South Pacific*, edited by Ton Otto and Nicholas Thomas, 65–90. Amsterdam: Harwood Academic Publishers.

Cochoy, Frank and Sophie Dubuisson-Quellier. 2013. 'The Sociology of Market Work'. *Economic Sociology: The European Electronic Newsletter* 15 (1): 4–11.

Collier, Stephen J, James Christopher Mizes and Antina Von Schnitzler. 2016. 'Preface: Public Infrastructures, Infrastructural Publics'. *Limn* 7. limn.it/articles/preface-public-infrastructures-infrastructural-publics/, accessed 8 July 2018.

Comms Update. 2013. 'Telikom Recruits Huawei for NBN Project'. 5 July. www.commsupdate.com/articles/2013/07/05/telikom-recruits-huawei-for-nbn-project/, accessed 28 November 2022.

Conley, John M and Cynthia A Williams. 2005. 'Engage, Embed, and Embellish: Theory Versus Practice in the Corporate Social Responsibility Movement'. *Journal of Corporation Law* 31: 1–38. doi.org/10.2139/ssrn.691521

Cookson, Robert. 2015. 'Digicel First Mobile Group to Block Ads in Battle Against Google'. *Financial Times,* 30 September.

Cox, John. 2011. 'Prosperity, Nation and Consumption: Fast Money Schemes in Papua New Guinea'. In *Managing Modernity in the Western Pacific*, edited by Mary Patterson and Martha Macintyre, 172–200. St Lucia: University of Queensland Press.

Cox, John. 2018. *Fast Money Schemes: Hope and Deception in Papua New Guinea*. Bloomington: Indiana University Press. doi.org/10.2307/j.ctv6mtfjm

Cranston, Belinda. 2013. 'Mobilising Politics in PNG'. ANU College of Asia & the Pacific. web.archive.org/web/20210327065710/http://asiapacific.anu.edu.au/news-events/all-stories/mobilising-politics-png, accessed 11 July 2023 (site discontinued).

Creaton, Siobahn. 2010. *A Mobile Fortune: The Life and Times of Denis O'Brien*. London: Aurum Press.

Cross, Jamie and Alice Street. 2009. 'Anthropology at the Bottom of the Pyramid'. *Anthropology Today* 25 (4): 4–9. doi.org/10.1111/j.1467-8322.2009.00675.x

Crowdy, Denis and Heather A Horst. 2022. 'We Just "SHAREit": Smartphones, Data and Music Sharing in Urban Papua New Guinea'. *TAJA: The Australian Journal of Anthropology* 33 (2): 247–62. doi.org/10.1111/taja.12444

Curry, George N, Elizabeth Dumu and Gina Koczberski. 2016. 'Bridging the Digital Divide: Everyday Use of Mobile Phones Among Market Sellers in Papua New Guinea'. In *Communicating, Networking: Interacting*, edited by Margaret E Robertson, 39–52. SpringerBriefs in Global Understanding. Cham: Springer. doi.org/10.1007/978-3-319-45471-9_5

Curwen, Peter, Jason Whalley and Pierre Vialle. 2019. *Disruptive Activity in a Regulated Industry: The Case of Telecommunications*. Bingley: Emerald Publishing. doi.org/10.1108/9781789734737

Deloitte Touche Tohmatsu. 2016. 'Why are Internet Prices High in Papua New Guinea?' Discussion Paper No. 148. Boroko, Papua New Guinea: The National Research Institute.

Digicel Foundation. 2018/19. *Annual Report*. Port Moresby.

Digicel Group Ltd. 2015. 'Form F-1, Registration Statement'. Washington DC: Securities and Exchange Commission.

Digicel PNG. 2012. 'NICTA Issues Content License to Digicel'. Facebook, 20 April, www.facebook.com/DigicelPNG/photos/a.400646567654/1015071130175 2655/?type=3, accessed 2 December 2022.

DiMaggio, Paul and Walter Powell. 1983. 'The Iron Cage Revisited: Institutional Isomorphism and Collective Rationality in Organizational Fields'. *American Sociological Review* 48 (2): 147–60. doi.org/10.2307/2095101

Di Nunzio, Marco. 2018. 'Anthropology of Infrastructure'. Governing Infrastructure Interfaces—Research Note 01 (June). London: London School of Economics and Political Science. lsecities.net/wp-content/uploads/2018/09/Governing-Infrastructure-Interfaces_Anthropology-of-infrastrcuture_MarcoDiNunzio.pdf, accessed 8 July 2019.

Diorio, Stephen. 2020. 'Proving the Financial Contribution of Sponsorships to the Business'. *Forbes*, 9 February. www.forbes.com/sites/forbesinsights/2020/02/09/proving-the-financial-contribution-of-sponsorships-to-the-business/?sh=4a526db53724, accessed 30 November 2022.

Dolan, Catherine and Dinah Rajak (eds). 2016. *The Anthropology of Corporate Social Responsibility*. New York: Berghahn. doi.org/10.2307/j.ctvgs09h2

Dolphin, Richard. 2003. 'Sponsorship: Perspectives on its Strategic Role'. *Corporate Communications* 8 (3): 173–86. doi.org/10.1108/13563280310487630

Donner, Jonathan. 2008. 'The Rules of Beeping: Exchanging Messages Via Intentional "Missed Calls" on Mobile Phones'. *Journal of Computer-Mediated Communication* 13 (1): 1–22. doi.org/10.1111/j.1083-6101.2007.00383.x

Donner, Jonathan. 2015. *After Access: Inclusion, Development, and a More Mobile Internet*. Cambridge, Massachusetts: MIT Press. doi.org/10.7551/mitpress/9740.001.0001

Donovan, Kevin P and Aaron K Martin. 2014. 'The Rise of African SIM Registration: The Emerging Dynamics of Regulatory Change'. *First Monday* 19, no. 2-3 (February). firstmonday.org/ojs/index.php/fm/article/view/4351/3820, accessed 13 December 2022.

Doron, Assa and Robin Jeffrey. 2013. *The Great Indian Phone Book: How the Cheap Cell Phone Changes Business, Politics and Daily Life*. Cambridge, Massachusetts: Harvard University Press. doi.org/10.4159/harvard.9780674074248

Duncan, Ronald. 2014. 'Telecommunications in Papua New Guinea'. In *Priorities and Pathways in Services Reform Part II — Political Economy Studies*, edited by Christopher Findlay, 27–44. World Scientific Publishing Co Pte Ltd. doi.org/10.1142/9789814504690_0002

Dwyer, Peter D and Monica Minnegal. 2020. 'COVID-19 and Facebook in Papua New Guinea: Fly River Forum'. *Asia & the Pacific Policy Studies* 7: 233–46. doi.org/10.1002/app5.312

Elapa, Jeffrey. 2019. 'PM Wants Review of Social Media'. *National*, 14 May. www. thenational.com.pg/pm-wants-review-of-social-media/, accessed 15 July 2019.

Errington, Frederick, Tatsuro Fujikura and Deborah Gewertz. 2012. 'Instant Noodles as an Antifriction Device: Making the BOP with PPP in PNG'. *American Anthropologist* 114 (1): 19–31. doi.org/10.1111/j.1548-1433.2011.01394.x

Fabian, Johannes. 1983. *Time and the Other: How Anthropology Makes Its Object*. New York: Columbia University Press.

Feld, Steven, Dennis Leonard, Jeremiah Ra Richards and Mickey Hart. 2019. *Voices of the Rainforest*. PhotoBook with audio CD and 7.1/5.1/2.0 film on BluRay. Papua New Guinea: VoxLox. www.voicesoftherainforest.org/, accessed 25 July 2023.

Ferguson, James. 2005. 'Seeing Like an Oil Company: Space, Security, and Global Capital in Neoliberal Africa'. *American Anthropologist* 107 (3): 377–82. doi.org/ 10.1525/aa.2005.107.3.377

Fiji Times. 2006. 'Digicel Enters PNG Market'. 12 September.

Forbes. n.d. 'Profile: Denis O'Brien'. www.forbes.com/profile/denis-obrien/?sh= 16ee1e6660ce, accessed 12 December 2022.

Foster, Robert J. 1993. 'Dangerous Circulation and Revelatory Display: Exchange Practices in a New Ireland Society'. In *Exchanging Products: Producing Exchange*, edited by Jane Fajans, 15–31. Oceania Monograph Series, no. 43. Sydney: University of Sydney.

Foster, Robert J. 1995. *Social Reproduction and History in Melanesia: Mortuary Ritual, Gift Exchange and Custom in the Tanga Islands*. Cambridge: Cambridge University Press.

Foster, Robert J. 1999. 'In God We Trust? The Legitimacy of Melanesian Currencies'. In *Money and Modernity: State and Local Currencies in Melanesia*, edited by Joel Robbins and David Akin, 214–31. Pittsburgh: University of Pittsburgh Press.

Foster, Robert J. 2002. *Materializing the Nation: Commodities, Consumption and Media in Papua New Guinea*. Bloomington: Indiana University Press.

Foster, Robert J. 2007. 'The Work of the "New Economy": Consumers, Brands and Value Creation'. *Cultural Anthropology* 22 (4): 707–31. doi.org/10.1525/ can.2007.22.4.707

Foster, Robert J. 2008. *Coca-Globalization: Following Soft Drinks from New York to New Guinea*. New York: Palgrave Macmillan.

Foster, Robert J. 2011. 'The Uses of Use Value: Marketing, Value Creation, and the Exigencies of Consumption Work'. In *Inside Marketing: Practices, Ideologies, Devices*, edited by Detlev Zwick and Julien Cayla, 42–57. Oxford: Oxford University Press. doi.org/10.1093/acprof:oso/9780199576746.003.0003

Foster, Robert J. 2013. 'Things to Do with Brands: Creating and Calculating Value'. In 'Value as Theory', edited by Ton Otto and Rane Willerslev, special issue, *HAU: Journal of Ethnographic Theory* 3 (1): 44–63. doi.org/10.14318/hau3.1.004

Foster, Robert J. 2014a. 'Adversaries into Partners? Brand Coca-Cola® and the Politics of Consumer-Citizenship'. In *Green Consumption: The Global Rise of Eco-Chic*, edited by Bart Barendregt and Rivke Jaffe, 19–36. London: Bloomsbury. doi.org/10.4324/9781003085508-3

Foster, Robert J. 2014b. 'Corporations as Partners: "Connected Capitalism" and The Coca-Cola Company'. *PoLAR: Political and Legal Anthropology Review* 37 (2): 246–58. doi.org/10.1111/plar.12073

Foster, Robert J. 2018. 'Top Up: The Moral Economy of Prepaid Mobile Phone Subscriptions'. In *The Moral Economy of Mobile Phones: Pacific Islands Perspectives,* edited by Robert J Foster and Heather A Horst, 107–25. Canberra: ANU Press. doi.org/10.22459/MEMP.05.2018.06

Foster, Robert J. 2020. 'The Politics of Media Infrastructure: Mobile Phones and Emergent Forms of Public Communication in Papua New Guinea'. *Oceania* 90 (1): 18–39. doi.org/10.1002/ocea.5241

Foster, Robert J. 2023. 'Tenuous Connectivity: Time, Citizenship, and Infrastructure in a Papua New Guinea Telecommunications Network'. *The Asia Pacific Journal of Anthropology* 24 (2): 91–115. doi.org/10.1080/14442213.2023.2177330

Foster, Robert J and Heather A Horst (eds). 2018. *The Moral Economy of Mobile Phones: Pacific Islands Perspectives.* Canberra: ANU Press. doi.org/10.22459/MEMP.05.2018

Fredericks, Rosalind and Mamadou Diouf. 2014. 'Introduction'. In *The Arts of Citizenship in African Cities: Infrastructures and Spaces of Belonging,* edited by Mamadou Diouf and Rosalind Fredericks, 1–23. New York: Palgrave Macmillan. doi.org/10.1057/9781137481887_1

Galgal, Kasek. 2017. 'Developing PNG's Cybercrime Policy: Local Contexts, Global Best Practice'. *The Interpreter,* 16 March. www.lowyinstitute.org/the-interpreter/developing-png-s-cybercrime-policy-local-contexts-global-best-practice, accessed 13 December 2022.

Garsten, Christina. 2004. 'Market Missions: Negotiating Bottom Line and Social Responsibility'. In *Market Matters: Exploring Cultural Processes in the Global Marketplace*, edited by Christina Garsten and Monica Lindh de Montoya, 69–90. Basingstoke: Palgrave.

Garsten, Christina. 2010. 'Ethnography at the Interface: "Corporate Social Responsibility" as an Anthropological Field of Inquiry'. In *Ethnographic Practice in the Present*, edited by Marit Melhuus, Jon P Mitchell and Helena Wulff, 56–67. New York: Berghahn. doi.org/10.1515/9780857455437-007

Garsten, Christina and Kerstin Jacobsson. 2007. 'Corporate Globalisation, Civil Society and Post-Political Regulation—Whither Democracy?' *Development Dialogue* 49: 143–57.

Gawi, Sylvester. 2019. 'Adopted from Communist China "PM Wants to Control Social Media"'. *Graun Blong Mi—My Land*, 13 May. sylvestergawi.blogspot.com/2019/05/adopted-from-communist-china-pm-wants.html, accessed 15 July 2019.

Genéve Internationale. n.d. 'Broadband Commission for Sustainable Development, ITU, UNESCO'. Who's Who, *Genéve Internationale*. www.geneve-int.ch/whoswho/broadband-commission-sustainable-development-itu-unesco, accessed 15 September 2022.

George, David. 2017. 'Former PNG Border Development Authority CEO Fred Konga Killed in "Execution Style" Shooting'. *Post-Courier*, 25 August.

Gilbert, Juliet. 2016. '"They're My Contacts, Not My Friends": Reconfiguring Affect and Aspirations Through Mobile Communication in Nigeria'. *Ethnos* 83 (2): 237–54. doi.org/10.1080/00141844.2015.1120762

Gillwald, Alison. 2015a. 'African Nations Use SIM Card Question to Mandate Control'. *BusinessDay*, 1 December. businessday.ng/technology/article/african-nations-use-sim-card-question-to-mandate-control/, accessed 26 July 2023.

Gillwald, Alison. 2015b. 'Comments for Stockholm Internet Forum (SIF 14)'. *Research ICT Africa*, 29 May. researchictafrica.net/2015/05/29/comments-for-stockholm-internet-forum-sif14-by-alison-gillwald/, accessed 26 July 2023.

González, Roberto J. 2020. *Connected: How a Mexican Village Built Its Own Cell Phone Network*. Berkeley: University of California Press. doi.org/10.1525/9780520975408

Good, Mary K. 2012. 'Modern Moralities, Moral Modernities: Ambivalence and Change Among Youth in Tonga'. PhD thesis, University of Arizona.

Gow, Gordon A and Jennifer Parisi. 2008. 'Pursuing the Anonymous User: Privacy Rights and Mandatory Registration of Prepaid Mobile Phones'. *Bulletin of Science, Technology & Society* 28 (1): 60–8. doi.org/10.1177/0270467607311487

Gregory, Chris A. 1982. *Gifts and Commodities.* New York: Academic Press.

Grigg, Angus. 2020. 'Huawei Data Centre Built to Spy on PNG'. *Australian Financial Review,* 11 August.

GSMA (Global System for Mobile Communications Association). 2014. *Connected Women—Striving and Surviving in Papua New Guinea: Exploring the Lives of Women at the Base of the Pyramid.* GSMA Connected Women. www.gsma.com/mobilefordevelopment/wp-content/uploads/2014/11/mWomen_PNG_v3.pdf, accessed 12 July 2023.

GSMA. 2016. *Mandatory Registration of Prepaid SIM Cards: Addressing Challenges Through Best Practices.* GSMA. www.gsma.com/publicpolicy/wp-content/uploads/2016/04/GSMA2016_Report_MandatoryRegistrationOfPrepaidSIM Cards.pdf, accessed 26 July 2023.

GSMA. 2021. *Informal Youth Employment in the Mobile Industry in Sub-Saharan Africa.* GSMA. www.gsma.com/mobilefordevelopment/wp-content/uploads/2021/12/YouthEmployment_R_WebSingles_2.pdf, accessed 26 July 2023.

GSMA. 2023. *The Mobile Economy Pacific Islands 2023.* GSMA. www.gsma.com/mobileeconomy/wp-content/uploads/2023/05/GSMA-ME-Pacific-Islands-2023.pdf, accessed 26 July 2023.

Gupta, Akhil. 2015. 'Suspension'. Theorizing the Contemporary, *Fieldsights,* 24 September. culanth.org/fieldsights/suspension, accessed 26 July 2023.

Gupta, Akhil. 2018. 'The Future in Ruins: Thoughts on the Temporality of Infrastructure'. In *The Promise of Infrastructure,* edited by Nikhil Anand, Akhil Gupta and Hannah Appel, 62–79. Durham: Duke University Press. doi.org/10.2307/j.ctv12101q3.6

Haihuie, Mark. 2017. 'Mobile Gambling Not Backed'. *National,* 14 March.

Handman, Courtney. 2013. 'Text Messaging in Tok Pisin: Etymologies and Orthographies in Cosmopolitan Papua New Guinea'. *Culture, Theory and Critique* 54 (3): 265–84. doi.org/10.1080/14735784.2013.818288

Harriman, Bethanie. 2018. 'PNG Minister Defends Controversial National Identification Program'. *ABC News,* 26 June, www.abc.net.au/news/2018-06-27/png-national-identification-program/9912590, accessed 26 July 2023.

Hart, Keith. 1986. 'Heads or Tails? Two Sides of the Coin'. *Man* 21 (4): 637–56. doi.org/10.2307/2802901

Harvey, Penelope, Casper Bruun Jensen and Atsuro Morita (eds). 2016. *Infrastructures and Social Complexity: A Companion.* New York: Routledge. doi.org/10.4324/9781315622880

Hempel, Jessi. 2018. 'What Happened to Facebook's Grand Plan to Wire the World?' Backchannel, *Wired,* 17 May. www.wired.com/story/what-happened-to-facebooks-grand-plan-to-wire-the-world/, accessed 13 December 2022.

Hendrie, Doug. 2011. 'Cops and Robbers Give Mobiles a Thumbs-up'. *The Australian,* 26 March.

Highet, Catherine, Michael Nique, Amanda HA Watson and Amber Wilson. 2019. *Digital Transformation: The Role of Mobile Technology in Papua New Guinea.* London, UK: GSMA.

Hobbis, Geoffrey. 2019. 'New Media, New Melanesia?' In *The Melanesian World,* edited by Eric Hirsch and Will Rollason, 546–60. New York: Routledge. doi.org/10.4324/9781315529691-33

Hobbis, Geoffrey. 2020. *The Digitizing Family: An Ethnography of Melanesian Smartphones.* New York: Palgrave Macmillan. doi.org/10.1007/978-3-030-34929-5

Horst, Heather A. 2006. 'The Blessings and Burdens of Communication: Cell Phones in Jamaican Transnational Social Fields'. *Global Networks* 6 (2): 143–59. doi.org/10.1111/j.1471-0374.2006.00138.x

Horst, Heather A. 2013. 'The Infrastructures of Mobile Media: Towards a Future Research Agenda'. *Mobile Media and Communication* 1 (1): 147–52. doi.org/10.1177/2050157912464490

Horst, Heather A. 2014. 'From Roots Culture to Sour Fruit: The Aesthetics of Mobile Branding Cultures in Jamaica'. *Visual Studies* 29 (2): 191–200. doi.org/10.1080/1472586X.2014.887272

Horst, Heather A. 2018. 'Creating Consumer-Citizens: Competition, Tradition, and the Moral Order of the Mobile Telecommunications Industry in Fiji'. In *The Moral Economy of Mobile Phones: Pacific Islands Perspectives,* edited by Robert J Foster and Heather A Horst, 73–92. Canberra: ANU Press. doi.org/10.22459/MEMP.05.2018.04

Horst, Heather A. 2021. 'The Anthropology of Mobile Phones'. In *Digital Anthropology*, edited by Haidy Geismar and Hannah Knox, 2nd ed., 65–84. New York: Routledge. doi.org/10.4324/9781003087885-6

Horst, Heather A and Robert J Foster. 2024. '5G and the Digital Imagination: Pacific Islands Perspectives from Fiji and Papua New Guinea'. *Media International Australia* 190 (1): 54–67. doi.org/10.1177/1329878X231199815

Horst, Heather A, Romitesh Kant and Eliki Drugunalevu. 2020. 'Smartphones and Parenting in Fiji: Regulation and Responsibility'. *Parenting for a Digital Future*, London School of Economics. blogs.lse.ac.uk/parenting4digitalfuture/2020/09/16/smartphones-and-parenting-in-fiji/, accessed 13 December 2022.

Horst, Heather A and Daniel Miller. 2005. 'From Kinship to Link-up: Cell Phones and Social Networking in Jamaica'. *Current Anthropology* 46 (5): 755–78. doi.org/10.1086/432650

Horst, Heather A and Daniel Miller. 2006. *The Cell Phone: An Anthropology of Communication*. New York: Berg.

Howell, Bronwyn E, Petrus H Potgieter and Ronald Sofe. 2019. 'Regulating for Telecommunications Competition in Developing Countries: Papua New Guinea'. *Asian-Pacific Economic Literature* 33 (1): 98–112. doi.org/10.1111/apel.12248

Howell, Catherine and Darrell M West. 2016. 'The Internet as a Human Right'. Brookings. www.brookings.edu/blog/techtank/2016/11/07/the-internet-as-a-human-right/, accessed 8 November 2022.

Howes, Stephen. 2021. 'Australia Buys Digicel, PNG's Mobile Monopoly'. *Devpolicyblog*, 26 October. devpolicy.org/australia-buys-digicel-pacific-pngs-mobile-monopoly-20211026/, accessed 8 November 2022.

Huang, Julia Qermezi. 2017. 'Digital Aspirations: "Wrong-Number" Mobile Phone Relationships and Experimental Ethics among Women Entrepreneurs in Rural Bangladesh'. *Journal of the Royal Anthropological Institute* 24 (1): 107–25. doi.org/10.1111/1467-9655.12754

International Finance Corporation. 2013. 'Digital Solar and PNG Solar Market Development'. International Finance Corporation. disclosures.ifc.org/project-detail/AS/594427/digicel-solar-and-png-solar-market-development, accessed 28 November 2022.

International Finance Corporation. n.d. 'About Us'. www.ifc.org/en/about, accessed 2 August 2023.

Irish Times. 2016. 'Cantillon: Blow for Denis O'Brien's Ad-blocking Campaign'. 6 September.

Islands Business. 2013. 'Electronic Wallet Making PNG Women Financially Independent'. 22 November.

ITU (International Telecommunication Union) n.d. 'The Journey to Universal and Meaningful Connectivity'. www.itu.int/itu-d/reports/statistics/2022/05/30/gcr-chapter-2/, accessed 15 September 2022.

ITU News. 2015. 'Leader Interview with Denis O'Brien'. *ITU News* (January/ February): 15–17. www.itu.int/bibar/ITUJournal/DocLibrary/ITU011-2015-01-en.pdf, accessed 10 November 2022.

James, David. 2018. 'Digicel Chairman Says Papua New Guinea "Fantastic" Location for Investment'. *Business Advantage PNG,* 18 June. www.business advantagepng.com/digicel-chairman-says-papua-new-guinea-fantastic-location-for-investment/, accessed 14 December 2022.

James, Paul, Yaso Nadarajah, Karen Haive, and Victoria Stead. 2012. *Sustainable Communities, Sustainable Development: Other Pathways for Papua New Guinea.* Honolulu: University of Hawai'i Press. doi.org/10.21313/hawaii/978082483 5880.001.0001

Jenkins, Henry. 2001. 'Convergence? I Diverge'. *MIT Technology Review,* June, 93.

Jorgensen, Dan. 2014. '*Gesfaia*: Mobile Phones, Phone Friends, and Anonymous Intimacy in Contemporary Papua New Guinea'. Paper presented at CASCA: Canadian Anthropology Society Conference. York University, Toronto, 30 April.

Jorgensen, Dan. 2018. 'Toby and "the Mobile System": Apocalypse and Salvation in Papua New Guinea's Wireless Network'. In *The Moral Economy of Mobile Phones: Pacific Islands Perspectives,* edited by Robert J Foster and Heather A Horst, 53–71. Canberra: ANU Press. doi.org/10.22459/MEMP.05.2018.03

Kachingwe, Nomsa and Alexandre Berthaud. 2014. 'Papua New Guinea: The Limits of the Mobile Payments Model'. Case Study 9. Universal Postal Union. www.findevgateway.org/case-study/2014/01/papua-new-guinea-limits-mobile-payments-model, accessed 14 December 2022.

Kalba, Kas. 2008. 'The Adoption of Mobile Phones in Emerging Markets: Global Diffusion and the Rural Challenge'. *International Journal of Communication* 2: 631–61.

Kant, Romitesh. 2022. 'Pacific Digital Toolbox Needed to Hammer out Misinformation'. *Radio New Zealand,* 19 July. www.rnz.co.nz/international/pacific-news/471243/pacific-digital-toolbox-needed-to-hammer-out-mis information, accessed 3 June 2023.

KCH (Kumul Consolidated Holdings). n.d. 'Kumul Consolidated Holdings'. Website. www.kch.com.pg/, accessed 14 July 2023.

Kenny, Erin. 2016. '"Phones Mean Lies": Secrets, Sexuality and the Subjectivity of Mobile Phones in Tanzania'. *Economic Anthropology* 3 (2): 254–65. doi.org/10.1002/sea2.12062

Ketterer Hobbis, Stephanie. 2018. 'Mobile Phones, Gender-Based Violence, and Distrust in State Services: Case Studies from Solomon Islands and Papua New Guinea'. *Asia Pacific Viewpoint* 59 (1): 60–73. doi.org/10.1111/apv.12178

Ketterer Hobbis, Stephanie and Geoffrey Hobbis. 2020. 'Non-/Human Infrastructures and Digital Gifts: The Cables, Waves and Brokers of Solomon Islands Internet'. *Ethnos* 87 (5): 851–73. doi.org/10.1080/00141844.2020.1828969

King, Phil. 2014. 'Tok Pisin and Mobail Teknoloji'. *Language and Linguistics in Melanesia* 32 (2): 118–52.

Kirsch, Stuart. 2014. *Mining Capitalism: The Relationship between Corporations and their Critics.* Berkeley: University of California Press. doi.org/10.1525/9780520957596

Klein, Naomi. 2007. *The Shock Doctrine: The Rise of Disaster Capitalism.* New York: Picador.

Knox, Hannah. 2017a. 'Affective Infrastructures and the Political Imagination'. *Public Culture* 29 (2): 363–84. doi.org/10.1215/08992363-3749105

Knox, Hannah. 2017b. 'An Infrastructural Approach to Digital Ethnography: Lessons from the Manchester Infrastructures of Social Change Project'. In *The Routledge Companion to Digital Ethnography*, edited by Larissa Hjorth, Heather Horst, Anne Galloway and Genevieve Bell, 354–62. New York: Routledge.

Knutson, Ryan and Sam Schechner. 2015. 'Is Facebook Friend or Foe for Telecom Operators?' *The Wall Street Journal*, 1 March.

Kraemer, Daniela 2015. '"Do You Have a Mobile?" Mobile Phone Practices and the Refashioning of Social Relationships in Port Vila Town'. *TAJA: The Australian Journal of Anthropology* 28 (1): 39–55. doi.org/10.1111/taja.12165

Kraemer, Daniela. 2018. '"Working the Mobile": Giving and Spending Phone Credit in Port Vila, Vanuatu'. In *The Moral Economy of Mobile Phones: Pacific Islands Perspectives,* edited by Robert J Foster and Heather A Horst, 93–106. Canberra: ANU Press. doi.org/10.22459/MEMP.05.2018.05

Kriem, Maya S. 2009. 'Mobile Telephony in Morocco: A Changing Sociality'. *Media, Culture & Society* 31 (4): 617–32. doi.org/10.1177/0163443709335729

Kumul Consolidated Holdings. 2015. 'New Satellite System Launched'. Kumul Consolidated Holdings, 22 October. www.kch.com.pg/new-satellite-system-launched/, accessed 10 February 2021.

Larkin, Brian. 2013. 'The Politics and Poetics of Infrastructure'. *Annual Review of Anthropology* 42: 327–43. doi.org/10.1146/annurev-anthro-092412-155522

Latour, Bruno and Peter Weibel (eds). 2005. *Making Things Public: Atmospheres of Democracy*. Cambridge, Massachusetts: MIT Press.

Laukai, Aloysius. 2008. 'Digicel Billboard Removed'. *National,* 7 January.

Lemanski, Charlotte. 2018. 'Infrastructural Citizenship: Spaces of Living in Cape Town, South Africa'. In *The Routledge Handbook on Spaces of Urban Politics,* edited by Kevin Ward, Andrew EG Jonas, Byron Miller and David Wilson, 350–60. London, UK: Routledge. doi.org/10.4324/9781315712468-35

Lemanski, Charlotte. 2022. 'Infrastructural Citizenship: Conceiving, Producing and Disciplining People and Place via Public Housing, from Cape Town to Stoke-on-Trent'. *Housing Studies* 37 (6): 932–54. doi.org/10.1080/02673037.2021.1966390

Ling, Rich and Heather A Horst. 2011. 'Mobile Communication in the Global South: An Introduction'. *New Media & Society,* 13 (3): 363–74. doi.org/10.1177/1461444810393899

Lipset, David. 2013. '*Mobail*: Moral Ambivalence and the Domestication of Mobile Telephones in Peri-Urban Papua New Guinea'. *Culture, Theory and Critique* 54 (3): 335–54. doi.org/10.1080/14735784.2013.826501

Lipset, David. 2017. 'Mobile Telephones as Public Sphere in Peri-urban Papua New Guinea'. *Le Journal de la Société des Océanistes* 144–145: 195–208. doi.org/10.4000/jso.7835

Lipset, David. 2018. 'A Handset Dangling in a Doorway: Mobile Phone Sharing in a Rural Sepik Village (Papua New Guinea)'. In *The Moral Economy of Mobile Phones: Pacific Islands Perspectives,* edited by Robert J Foster and Heather A Horst, 19–37. Canberra: ANU Press. doi.org/10.22459/MEMP.05.2018.01

Little, Christopher Albert John Lonc. 2016. 'The Precarity of Men: Youth, Masculinity, and Money in a Papua New Guinean Town'. PhD thesis, University of Toronto.

Logan, Sarah. 2012. 'Rausim! Digital Politics in Papua New Guinea'. SSGM Discussion Paper 2012/9. dpa.bellschool.anu.edu.au/sites/default/files/publications/attachments/2015-12/2012_9_0.pdf, accessed 9 July 2019.

Logan, Sarah and Miranda Forsyth. 2018. 'Access All Areas? Telecommunications and Human Rights in Papua New Guinea'. *Human Rights Defender* 27 (3): 18–20.

Logan, Sarah and Joseph Suwamaru. 2017. 'Land of the Disconnected: A History of the Internet in Papua New Guinea'. In *The Routledge Companion to Global Internet Histories,* edited by Gerard Goggin and Mark McLelland, 284–95. New York: Routledge. doi.org/10.4324/9781315748962-20

Loop PNG. 2014. 'Remote School in Morobe Receives K75,000'. 25 July.

Loop PNG. 2015. 'From the Mountains to the Sea'. 18 February.

Loop PNG. 2016. 'Firm Aims to Increase Data Penetration'. 28 July. www.looppng. com/content/firm-aims-increase-data-penetration, accessed 14 December 2022.

Loop PNG. 2020. 'Digicel Steps Out as a Digital Operator'. 23 October. www.looppng.com/png-news/digicel-steps-out-digital-operator-95312, accessed 14 December 2022.

MacCarthy, Michelle. 2011. 'Technologies and Economies of Mobility: The Introduction of Mobile Phones in the Trobriand Islands'. Paper presented at the Australian Anthropological Society Annual Conference, Perth, WA.

Macdonald, Fraser and Jonathan Kirami. 2015. 'Women, Mobile Phones, and M16s: Contemporary New Guinea Highlands Warfare'. *TAJA: The Australian Journal of Anthropology* 28 (1): 104–19. doi.org/10.1111/taja.12175

Malinowski, Bronislaw. 1922. *Argonauts of the Western Pacific: An Account of Native Enterprise and Adventure in the Archipelagoes of Melanesian New Guinea.* New York: E.P. Dutton.

Mantz, Jeffrey W. 2018. 'From Digital Divides to Creative Destruction: Epistemological Encounters in the Regulation of the "Blood Mineral" Trade in the Congo'. *Anthropological Quarterly* 91 (2): 525–50. doi.org/10.1353/anq. 2018.0025

Martin, Keir. 2013. *The Death of the Big Men and the Rise of the Big Shots: Custom and Conflict in East New Britain.* New York: Berghahn.

Mas, Ignacio and John Staley. 2014. 'Why Equity Bank Felt It Had to Become a Telco – Reluctantly'. Consultative Group to Assist the Poor (CGAP), 18 June. www.cgap.org/blog/why-equity-bank-felt-it-had-become-telco-reluctantly, accessed 14 December 2022.

Mauss, Marcel. 1923–24. 'Essai sur le Don: Forme et raison de l'échange dans les sociétés primitives'. *l'Année Sociologique,* seconde série, 1923–1924.

McGarry, Dan. 2020. 'Vanuatu Pioneers Digital Cash as Disaster Relief'. *The Guardian*, 6 November.

McLeod, Shane. 2020. 'Debt Threatens Digicel's Pacific Dominance'. *The Interpreter*, The Lowy Institute, 22 June. www.lowyinstitute.org/the-interpreter/debt-threatens-digicel-s-pacific-dominance, accessed 14 December 2022.

Meese, James, Rowan Wilken and Ioana Chan Mow. 2019. 'Uneven Topologies of Communication: Mobiles and Transnational Location in Samoa'. In *Location Technologies in International Context,* edited by Rowan Wilken, Gerard Goggin and Heather Horst, 93–107. New York: Routledge. doi.org/10.4324/97813 15544823-8

Metcalf, Tom. 2013. 'Denis O'Brien, Ireland's Version of Mexico's Carlos Slim'. *Washington Post,* 13 November.

Miller, Daniel. 2006. 'The Unpredictable Mobile Phone'. *BT Technology Journal* 24 (3): 41–48. doi.org/10.1007/s10550-006-0074-1

Miller, Daniel, Laila Abed Rabho, Patrick Awondo, Maya de Vries, Marília Duque, Pauline Garvey, Laura Haapio-Kirk, Charlotte Hawkins, Alfonso Otaegui, Shireen Walton and Xinyuan Wang. 2021. *The Global Smartphone: Beyond a Youth Technology.* London: UCL Press. doi.org/10.14324/111.9781787359611

Miller, Vincent. 2011. *Understanding Digital Culture.* London: Sage Publications.

Mintz, Sidney W. 1985. *Sweetness and Power: The Place of Sugar in Modern History.* New York: Viking.

Morofa, Delly. 2014. 'Kruak: Pilot Research Successfully Conducted'. *PNG Office of Censorship Newsletter* 2 (April–June).

Munn, Nancy D. 1986. *The Fame of Gawa: A Symbolic Study of Value Transformation in a Massim (Papua New Guinea) Society.* Cambridge: Cambridge University Press.

Munoz, Gabriella. 2020. 'Are Satellites the Future of Affordable Internet in Papua New Guinea?' *Business Advantage PNG*, 16 March.

Murdock, Ryan. 2022. 'Disconnected: Electrification in Papua New Guinea'. *Harvard International Review,* 16 May. hir.harvard.edu/electrification-in-papua-new-guinea/, accessed 9 November 2022.

National. 2007a. 'PANGTEL Restrained over Digicel Licence'. 26 July.

National. 2007b. 'Letters'. 31 July.

National. 2007c. 'World Bank Backs Digicel'. 31 July.

National. 2007d. 'Telikom Setting the Record Straight'. 27 July.

National. 2007e. 'Telikom's New Image'. 12 September.

National. 2007f. 'The Vandals Are Always There'. 27 September.

National. 2009. 'Gaming Board Zeroes in on Digicel Promotions'. 3 April.

National. 2010a. 'Landowners Call on Digicel to Remove Towers'. 10 November.

National. 2010b. 'Digicel Tower Vandalised'. 7 February.

National. 2011. 'Digicel Pays K58,000 Land Compo'. 17 January.

National. 2013. 'Kainantu Schools Disappointed'. 7 October.

National. 2014a. 'Digicel Backs City Park'. 25 July.

National. 2014b. 'Digicel Set for New Bio System'. 19 June.

National. 2015a. 'Company: Games a Great Way to Showcase Services'. 28 July.

National. 2015b. 'Clinics Providing Tips for Smartphone Customers'. 28 May.

National. 2015c. 'Media Firm Launches New APP'. 16 November.

National. 2015d. 'Don't Blame Parents for Internet Abuse'. 24 July.

National. 2017a. 'Digicel Gaming Services a Worry for NICTA'. 15 June.

National. 2017b. 'Digicel PNG Launches My Digicel Application'. 20 January.

National. 2020a. 'Digicel Invested Kina 3.4 Billion in Communities, Says Official'. 1 October.

National. 2020b. 'Bmobile Urged to Reach Out'. 27 July.

National. 2020c. 'Huawei Preferred Comms Equipment Vendor, Says Muthuvel'. 13 August.

National. 2020d. 'Muthuvel Aims to Lift Standard'. 14 August.

National. 2020e. 'Digicel, Coca-Cola Team Up in New CellMoni Promo'. 1 October.

National. 2021. 'Telstra Will Continue Digicel Foundation Ops'. 26 October.

National. 2023. 'Users Urged to Upgrade SIMs'. 12 January.

National Parliament of Papua New Guinea. 2013. Hansard. National ICT Policy, Ministerial Statement, 13 November. www.parliament.gov.pg/uploads/hansard/H-09-20131113-M10-D02.pdf, accessed 26 July 2023.

Network Strategies. 2013. 'Affordability of Mobile Services Hampered by Quasi-Monopolies in the Pacific', 12 March. strategies.nzl.com/industry-comment/affordability-of-mobile-services-hampered-by-quasi-monopolies-in-the-pacific/, accessed 25 November 2022.

Newens, Chris. 2021. 'Papua New Guinea Calling'. *Rest of World,* 5 January. restofworld.org/2021/papua-new-guinea-calling/, accessed 27 November 2022.

O'Brien, Denis. 2020. 'Digicel: Continuing Tradition of Direct Participation in Emergency Response'. Broadband Commission for Sustainable Development, 21 September. web.archive.org/web/20230131170429/https://www.broadbandcommission.org/insight/digicel-emergency-response/, accessed 12 July 2023.

O'Brien, Denis. n.d. 'Patron's Message'. Digicel Foundation. www.digicelfoundation.org/png/en/home/about/patrons-message.html, accessed 6 December 2022.

'Ofa, Siope V. 2011. 'Telecommincation Regulatory Reforms and the Credibility Problem: Case Studies from Papua New Guinea and Tonga'. In *The Political Economy of Economic Reform in the Pacific,* edited by Ron Duncan, 63–93. Manila: Asian Development Bank.

'Ofa, Siope V. 2012. *Telecommunications Regulatory Reform in Small Island Developing States: The Impact of WTO's Telecommunications Commitment.* Newcastle-Upon-Tyne: Cambridge Scholars Publishing.

ONE Papua New Guinea. 2017. 'Digicel Mobile Gambling Services Compliant: NGCB'. 10 April. www.onepng.com/2017/04/digicel-mobile-gambling-services.html, accessed 30 November 2022.

Oxford Business Group. 2012a. 'On It Grows: Little Sign of a Slowdown as Providers Diversify Products and Services in Line with Market Demand'. In *The Report: Papua New Guinea 2012,* Oxford Business Group. oxfordbusinessgroup.com/overview/it-grows-little-sign-slowdown-providers-diversify-products-and-services-line-market-demand, accessed 14 December 2022.

Oxford Business Group. 2012b. 'Papua New Guinea: ICT Communication Breakdown'. In *The Report: Papua New Guinea 2012,* Oxford Business Group. oxfordbusinessgroup.com/news/papua-new-guinea-ict-communication-breakdown, accessed 28 November 2022.

Oxford Business Group. 2012c. 'Banking on Mobiles: Operators are Rolling Out New Services, But May Face a Shallow Market'. In *The Report: Papua New Guinea 2012*, Oxford Business Group. oxfordbusinessgroup.com/analysis/banking-mobiles-operators-are-rolling-out-new-services-may-face-shallow-market, accessed 14 December 2022.

Oxford Business Group. 2014. 'On the Rise: Despite the Challenges Facing the Sector, Increased Competition and Investment are Fueling Growth'. In *The Report: Papua New Guinea 2014*, Oxford Business Group. oxfordbusinessgroup.com/reports/papua-new-guinea/2014-report/economy/despite-the-challenges-facing-the-sector-increased-competition-and-investment-are-fuelling-growth, accessed 14 December 2022.

Oxford Business Group. 2015. 'Bhanu Sud, CEO, EMTV: Interview'. In *The Report: Papua New Guinea 2015*, Oxford Business Group.

Oxford Business Group. 2016. 'Cybercrime Laws Being Updated in Papua New Guinea'. In *The Report: Papua New Guinea 2016*, Oxford Business Group.

Oxford Business Group. 2019. 'Infrastructure Investments and Increased Competition to Support Burgeoning Digital Economy in Papua New Guinea'. In *The Report: Papua New Guinea 2019*, Oxford Business Group.

Oxford Business Group. 2020. 'What is Driving Enhanced ICT Services in Papua New Guinea?' In *The Report: Papua New Guinea 2020*, Oxford Business Group.

Pacific Islands Report. 2008. 'Disgruntled PNG Landowners Attack Digicel Towers'. 26 November.

Pacific Media Centre. 2015. 'PNG: Former PM Criticises New Cybercrime Laws 2015'. 29 October. pmc.aut.ac.nz/pacific-media-watch/png-former-pm-criticises-new-cybercrime-laws-9467, accessed 14 December 2022.

Parks, Lisa and Nicole Starosielski. 2015. 'Introduction'. In *Signal Traffic: Critical Studies of Media Infrastructures*, edited by Lisa Parks and Nicole Starosielski, 1–27. Champaign: University of Illinois Press. doi.org/10.5406/illinois/9780252039362.001.0001

Paul, Mark. 2015. 'Regulators Warn Digicel Over Ad-blocking Google and Facebook'. *Irish Times,* 4 November.

Peebles, Gustav. 2014. 'Rehabilitating the Hoard: The Social Dynamics of Unbanking in Africa and Beyond'. *Africa* 84 (4): 595–613. doi.org/10.1017/S0001972014000485

People's Daily Online. 2009. 'Chinese Telco Giant Huawei to Invest in PNG'. 22 April. web.archive.org/web/20220223175738/en.people.cn/90001/90778/90857/90861/6642611.html, accessed 12 July 2023.

Peseckas, Ryan. 2014. 'Island Connections: Mobile Phones and Social Change in Rural Fiji'. PhD thesis, University of Florida.

Plantin, Jean-Christophe and Aswin Punathambekar. 2019. 'Digital Media Infrastructures: Pipes, Platforms, and Politics'. *Media, Culture & Society* 41 (2): 163–74. doi.org/10.1177/0163443718818376

PNG Attitude. 2009. 'Morauta Puts No Confidence Back on Agenda'. *Keith Jackson & Friends: PNG Attitude*, November. www.pngattitude.com/2009/11/morauta-puts-no-confidence-back-on-agenda.html, accessed 16 September 2022 (site discontinued).

PNG Attitude. 2018. 'Digicel Needs to Come Clean with Illiterate Landowners'. *Keith Jackson & Friends: PNG Attitude*, 23 December. web.archive.org/web/20181223003715/asopa.typepad.com/asopa_people/2018/12/digicel-needs-to-come-clean-with-poor-illiterate-landowners.html, accessed 12 July 2023.

PNGBLOGS. 2015. 'Fred Konga Stupidly Tries to Intimidate Corruption Fighters'. 14 August. web.archive.org/web/20221002054130/www.pngblogs.com/2015/08/fred-konga-stupidly-tries-to-intimidate.html, accessed 12 July 2023.

PNG Bulletin Online. 2021. 'PM Announces Telikom and BMobile Merger, KTHL Abolished'. 18 September, thepngbulletin.com/news/business/pm-announces-telikom-and-bmobile-merger-kthl-abolished/, accessed 14 December 2022.

PNG DataCo. n.d. 'About Us'. www.pngdataco.com/about/, accessed 26 July 2023.

Post-Courier. 2006a. 'Telikom Service Shaken'. 3 February.

Post-Courier. 2006b. 'Mobile Bidder'. 4 September.

Post-Courier. 2006c. 'No Board at PANGTEL'. 10 October.

Post-Courier. 2007a. 'Competition in All Sectors'. 5 July.

Post-Courier. 2007b. 'PANGTEL Says Report Not True'. 11 July.

Post-Courier. 2007c. 'Digicel Hits Telikom over Competition'. 28 June.

Post-Courier. 2007d. 'Security Fears with Separate Network'. 13 July.

Post-Courier. 2007e. 'Mobile Comp Heats Up'. 22 May.

Post-Courier. 2007f. 'Digicel to Hire only the Best'. 29 June.

Post-Courier. 2007g. 'Chief Pushes Illegal Line'. 2 August.

Post-Courier. 2007h. 'Digicel Licence Revoked'. 25 July.

Post-Courier. 2007i. 'Concerns over Govt's Handling of Digicel Affair'. 3 September.

Post-Courier. 2007j. 'ICCC Tells Telikom to Carry On'. 30 May.

Post-Courier. 2007k. 'Talks on the Horizon'. 3 September.

Post-Courier. 2007l. Letter to the Editor. 30 July.

Post-Courier. 2010. 'Outcry Stops Papua New Guinea Phone Lottery'. 8 November.

Post-Courier. 2012a. Advertisement. 4 May.

Post-Courier. 2012b. 'Digicel, Please Make Winners Known to the Public'. Letter to the Editor. 27 April.

Post-Courier. 2012c. 'All Out Brawl on "Islands of Love"'. 26 April.

Post-Courier. 2012d. 'Mobile Phones a Security Threat'. 11 October.

Post-Courier. 2012e. 'MP Moves to Have Mobile Phones Registered for Tracking'. 31 December.

Post-Courier. 2013a. 'MP's Call to Register SIMs Laudable'. 19 March.

Post-Courier. 2013b. 'SIM Card Registration to Control Users'. 17 November.

Post-Courier. 2015a. 'RESTORE SERVICE'. Letter to the Editor. 6 July.

Post-Courier. 2015b. 'Smartphone Sales Through the Roof'. 13 July.

Post-Courier. 2015c. 'Digicel Upgrades Network'. 29 June.

Post-Courier. 2015d. 'Phone Texting Affects Spelling'. 10 August.

Post-Courier. 2017a. 'Vandalism to Towers Costly, Says Digicel Chief'. 14 July.

Post-Courier. 2017b. 'Goroka Mobile Users Line Up to Register'. 29 December.

Post-Courier. 2017c. 'BIMA Pays 6mil'. 22 February.

Post-Courier. 2018a. 'Foreign Company "Disappears Without a Trace"'. 1 February.

Post-Courier. 2018b. 'PNGSDP Allocates K105 Million for Telecommunication Upgrade in Western'. 19 September.

Post-Courier. 2019. '35,000 Indirect Incomes Rely on Network'. 9 August.

Post-Courier. 2020. 'Digicel Cell Moni Wallet Awareness Drive Hits the Road'. 7 January.

Post-Courier. 2022a. 'Tower Sharing Costly'. 2 May.

Post-Courier. 2022b. 'Telstra Takes Over Digicel'. 15 July.

Potter, Robert. 2021. 'Papua New Guinea and China's Debt Squeeze'. *The Diplomat*, 2 February. thediplomat.com/2021/02/papua-new-guinea-and-chinas-debt-squeeze/, accessed 15 December 2022.

Prahalad, CK. 2010. *The Fortune at the Bottom of the Pyramid: Eradicating Poverty Through Profits*. Rev. ed. Upper Saddle River, NJ: Prentice Hall.

Punaha, Charles. 2007. 'PANGTEL Refutes Claims of Revoking Digicel's Interim Spectrum License on Ministerial Directives'. *National*, 26 July.

Pype, Katrien. 2021. '(Not) In Sync—Digital Time and Forms of (Dis-)connecting: Ethnographic Notes from Kinshasa (DR Congo)'. *Media, Culture & Society* 43 (7): 1197–212. doi.org/10.1177/0163443719867854

Radio New Zealand. 2016. 'Fears of Overreach with New PNG Cyber Crime Law'. 15 August. www.rnz.co.nz/international/programmes/datelinepacific/audio/201812187/fears-of-overreach-with-new-png-cyber-crime-law, accessed 15 December 2022.

Radio New Zealand. 2023. 'Deaths in Tribal Fighting in PNG's Porgera District'. 30 April. www.rnz.co.nz/international/pacific-news/488883/deaths-in-tribal-fighting-in-png-s-porgera-district, accessed 26 July 2023.

Rajak, Dinah. 2011a. 'Theatres of Virtue: Collaboration, Consensus, and the Social Life of Corporate Social Responsibility'. *Focaal–Journal of Global and Historical Anthropology* 60: 9–20.

Rajak, Dinah. 2011b. *In Good Company: An Anatomy of Corporate Social Responsibility*. Stanford: Stanford University Press. doi.org/10.1515/9780804781619

Robb, Alice. 2014. 'In this Papua New Guinea Village, People Use Cell Phones to Call the Dead'. *The New Republic*, 17 June. newrepublic.com/article/118216/cell-phones-papua-new-guinea-used-call-dead-people, accessed 15 December 2022.

Rooney, Michelle Nayahamui. 2012. 'Can Social Media Transform Papua New Guinea? Reflections and Questions'. *Devpolicyblog*, 31 July. devpolicy.org/can-social-media-transform-papua-new-guinea-reflections-and-questions20120731/, accessed 15 December 2022.

Rooney, Michelle Nayahamui, Martin Davies and Stephen Howes. 2020. 'Mi Gat Y: Is Digicel PNG's Loan Scheme Predatory?' *Devpolicyblog*, 25 May. devpolicy.org/mi-gat-y-is-digicel-pngs-loan-scheme-predatory-20200521/, accessed 15 December 2022.

Roy, Eleanor Ainge. 2018. 'Papua New Guinea Bans Facebook for a Month to Root Out "Fake Users"'. *Guardian,* 29 May. www.theguardian.com/world/2018/may/29/papua-new-guinea-facebook-ban-study-fake-users, accessed 15 December 2022.

Sahlins, Marshall D. 1963. 'Poor Man, Rich Man, Big-Man, Chief: Political Types in Melanesia and Polynesia'. *Comparative Studies in Society and History* 5 (3): 285–303. doi.org/10.1017/S0010417500001729

Sahlins, Marshall. 2011. 'What Kinship Is (Part One)'. *Journal of the Royal Anthropological Institute* 17 (1): 2–19. doi.org/10.1111/j.1467-9655.2010.01666.x

Sahlins, Marshall. 2012. 'Alterity and Autochthony: Austronesian Cosmographies of the Marvelous'. *HAU: Journal of Ethnographic Theory* 2 (1): 131–60. doi.org/10.14318/hau2.1.008

Schieffelin, Edward L. 1976. *The Sorrow of the Lonely and the Burning of the Dancers.* New York: St Martin's Press.

Shamir, Ronen. 2008. 'The Age of Responsibilization: On Market-Embedded Morality'. *Economy and Society* 37 (1): 1–19. doi.org/10.1080/03085140701760833

Shamir, Ronen. 2010. 'Capitalism, Governance, and Authority: The Case of Corporate Social Responsibility'. *Annual Review of Law and Social Science* 6: 531–53. doi.org/10.1146/annurev-lawsocsci-102209-153000

Sharp, Timothy Lachlan. 2012. 'Following Buai: The Highlands Betel Nut Trade, Papua New Guinea'. PhD thesis, The Australian National University.

Shever, Elana. 2010. 'Engendering the Company: Corporate Personhood and the "Face" of an Oil Company in Metropolitan Buenos Aires'. *PoLAR: Political and Legal Anthropology Review* 33 (1): 26–46. doi.org/10.1111/j.1555-2934.2010.01091.x

Shever, Elana. 2012. *Resources for Reform: Oil and Neoliberalism in Argentina.* Stanford: Stanford University Press. doi.org/10.1515/9780804783200

Silverstone, Roger. 1995. 'Convergence is a Dangerous Word'. *Convergence* 1 (1): 11–13. doi.org/10.1177/135485659500100102

Simone, AbdouMaliq. 2004. 'People as Infrastructure: Intersecting Fragments in Johannesburg'. *Public Culture* 16 (3): 407–29. doi.org/10.1215/08992363-16-3-407

Simone, AbdouMaliq. 2012. 'Infrastructure: Commentary by AbdouMaliq Simone'. *Cultural Anthropology.* journal.culanth.org/index.php/ca/infrastructure-abdoumaliq-simone, accessed 8 July 2019.

Sinanan, Jolynna, Heather Horst and Romitesh Kant. 2022. 'Infrastructuring in the Global South: Ethnographic Perspectives on Tourism, Media and Development'. In *The SAGE Handbook of the Digital Media Economy*, edited by Terry Flew, Jennifer Holt and Julian Thomas, 145–69. London: Sage. doi.org/10.4135/9781529757170.n10

Sinclair, James Patrick. 1984. *Uniting a Nation: The Postal and Telecommunication Services of Papua New Guinea*. Melbourne; New York: Oxford University Press.

Solomon Times. 2008. 'Digicel is Coming to Town'. 9 September.

Sora, Mihai and Jonathan Pryke. 2021. 'Telstra's Digicel Pacific Challenge'. Lowy Institute, 29 October. www.lowyinstitute.org/publications/telstra-s-digicel-pacific-challenge, accessed 9 November 2022.

Starosielski, Nicole. 2015. *The Undersea Network*. Durham: Duke University Press. doi.org/10.2307/j.ctv11smhj2

Strathern, Marilyn. 1991. *Partial Connections*. Savage, Maryland: Rowman & Littlefield Publishers.

Sullivan, Nancy. 2010a. *Revised Social Assessment for the PNG Rural Communications Project*. Washington DC: World Bank.

Sullivan, Nancy. 2010b. 'Fieldwork Report in Support of an Environmental and Social Management Framework, for the World Bank supported Rural Communications Fund Project in East Sepik and Simbu Provinces, Papua New Guinea'. www.academia.edu/4492654/Fiedlwork_Report_in_Support_of_an_Envrionmental_and_Social_Management_Framework_for_Rural_Communications, accessed 12 July 2023.

Suwamaru, Joseph. 2015. 'Aspects of Mobile Phone Usage for Socioeconomic Development in Papua New Guinea'. SSGM Discussion Paper 2015/11. Canberra: The Australian National University. dpa.bellschool.anu.edu.au/sites/default/files/publications/attachments/2016-07/dp_2015_11_suwamaru_proof2.pdf, accessed 26 July 2023.

Suwamaru, Joseph. 2020. 'Beneath the Veil of the Kumul Submarine Cable Network'. *Electronic Journal of Informatics* 2: 1–25.

SWRVE. 2020. 'Digicel Triples Engagement and Increases Retention by Over 50%'. Case Study. www.gsma.com/membership/wp-content/uploads/2020/07/Swrve-Digicel-Case-Study-A4.pdf, accessed 15 February 2024.

SWRVE. n.d. 'About the Company'. www.swrve.com/company, accessed 23 July 2023.

Tabureguci, Dionesia. 2010. 'IFC's US$140M Splash in BSP, is there a Conflict of Interest?' *Islands Business,* 31 August, reposted at PNGBLOGS. web.archive.org/web/20221128014543/https://www.pngblogs.com/2010/08/ifcs-us140m-splash-in-bsp-is-there.html, accessed 12 July 2023.

Taylor, Erin and Heather Horst. 2018. 'Banking on the Market: Mobile Phones and Social Goods Provision in Haiti'. *Anthropological Quarterly* 91 (2): 551–80. doi.org/10.1353/anq.2018.0026

Taylor, John P. 2015. 'Drinking Money and Pulling Women: Mobile Phone Talk, Gender and Agency in Vanuatu'. *Anthropological Forum,* 25: 1–16. doi.org/10.1080/00664677.2015.1071238

Telban, Borut and Daniela Vávrová. 2014. 'Ringing the Living and the Dead: Mobile Phones in a Sepik Society'. *TAJA: The Australian Journal of Anthropology* 25 (2): 223–38. doi.org/10.1111/taja.12090

Telecompaper. 2021. 'Bmobile Completes 4G Upgrade'. 18 January. www.telecompaper.com/news/bmobile-completes-4g-upgrade--1368915, accessed 2 June 2023.

Telepin Software Systems Inc. n.d. 'Digicel Case Study: Mobile Wallet Helps Reach Unbanked and Underbanked Customers'. www.telepin.com/case-study/mobile-wallet-helps-reach-unbanked-and-underbanked-customers/, accessed 17 February 2020.

Telikom PNG Limited. 2012. 'Submission to Public Inquiry Declaration of Wholesale Services in International Connectivity Markets'. 19 October. www.nicta.gov.pg/international-connectivity-markets/, accessed 12 July 2023.

Temple, Olga. 2016. 'The Influence of Mobile Phones on the Languages and Cultures of Papua New Guinea'. In *Indigenous Peoples and Mobile Technologies,* edited by Laurel Evelyn Dyson, Stephen Grant and Max Hendriks, 274–92. London: Routledge.

Temple, Olga, Alopi Apakali, Dorothy Bai, Lilly John, Gabriel Matiwat and Malinda Ginmauli. 2009. *PNG SMS Serendipity, or sms@upng.ac.pg (aka student texting lingo).* Port Moresby: University of Papua New Guinea Press.

Tenhunen, Sirpa. 2018. *A Village Goes Mobile: Telephony, Mediation and Social Change in Rural India.* New York: Oxford University Press. doi.org/10.1093/oso/9780190630270.001.0001

Thomas, Verena, Jackie Kauli, Wendy Bai Magea, Robert Foster and Heather Horst. 2018. *Mobail Goroka.* Video. Center for Social and Creative Media, University of Goroka. www.youtube.com/watch?v=qYuPxueHGoU, accessed 26 July 2023.

Thompson, E.P. 1971. 'The Moral Economy of the English Crowd in the Eighteenth Century'. *Past and Present* 50: 71–136. doi.org/10.1093/past/50.1.76

Titifanue, Jason, Jope Tarai, Romitesh Kant and Glen Finau. 2016. 'From Social Networking to Activism: The Role of Social Media in the Free West Papua Campaign'. *Pacific Studies* 39 (3): 255–80.

Titus, Asha Susan. 2019. 'Mobile Phones and the Promises of Connectivity: Interrogating the Role of Information, Communication Technologies (ICTs) in Marketisation'. MA thesis, The Australian National University.

Uimonen, Paula. 2015. '"Number Not Reachable": Mobile Infrastructure and Global Racial Hierarchy in Africa'. *Journal des Anthropologues* 142–143: 29–47. doi.org/10.4000/jda.6197

United Nations Development Programme. 2019. 'Human Development Reports, Literacy Rate, Adult (% Ages 15 and Older)'. hdr.undp.org/en/indicators/101406, accessed 20 January 2020.

Venkatesan, Soumyha, Laura Bear, Penny Harvey, Sian Lazar, Laura Rival and AbdouMaliq Simone. 2018. 'Attention to Infrastructure Offers a Welcome Reconfiguration of Anthropological Approaches to the Political'. *Critique of Anthropology* 38 (1): 3–52. doi.org/10.1177/0308275X16683023

Vokes, Richard. 2018. 'Before the Call: Mobile Phones, Exchange Relations, and Social Change in South-western Uganda'. *Ethnos* 83 (2): 274–90. doi.org/10.1080/00141844.2015.1133689

von Schnitzler, Antina. 2008. 'Citizenship Prepaid: Water, Calculability and Techno-Politics in South Africa'. *Journal of Southern African Studies* 34 (4): 899–917. doi.org/10.1080/03057070802456821

Wafer, Alex. 2012. 'Discourses of Infrastructure and Citizenship in Post-Apartheid Soweto'. *Urban Forum* 23: 233–43. doi.org/10.1007/s12132-012-9146-0

Wallis, Cara. 2011. 'Mobile Phones Without Guarantees: The Promises of Technology and the Contingencies of Culture'. *New Media & Society* 13 (3): 471–85. doi.org/10.1177/1461444810393904

Wallis, Cara. 2013. *Technomobility in China: Young Migrant Women and Mobile Phones*. New York: New York University Press.

Wardlow, Holly. 2018. 'HIV, Phone Friends, and Affective Technology in Papua New Guinea'. In *The Moral Economy of Mobile Phones: Pacific Islands Perspectives*, edited by Robert J Foster and Heather A Horst, 39–52. Canberra: ANU Press. doi.org/10.22459/MEMP.05.2018.02

Watson, Amanda HA. 2011. 'The Mobile Phone: The New Communication Drum of Papua New Guinea'. PhD thesis, Queensland University of Technology.

Watson, Amanda HA. 2012. 'Tsunami Alert: The Mobile Phone Difference'. *The Australian Journal of Emergency Management* 27 (4): 44–48.

Watson, Amanda HA. 2017. 'Does PNG Rank Highly for Internet Porn Searches?' *Devpolicyblog*, 31 January. devpolicy.org/does-png-rank-highly-internet-porn-searches-20170131/, accessed 20 December 2022.

Watson, Amanda HA. 2018. 'Compulsory SIM Card Registration in Papua New Guinea'. *Devpolicyblog*, 24 January. www.devpolicy.org/compulsory-sim-card-registration-in-png-20180124/, accessed 15 July 2019.

Watson, Amanda HA. 2020a. 'Internet Prices in Papua New Guinea'. *Devpolicyblog*, 30 January. devpolicy.org/internet-prices-in-papua-new-guinea-20200130/, accessed 20 December 2022.

Watson, Amanda HA. 2020b. 'Mobile Phone Registration in Papua New Guinea: Will the Benefits Outweigh the Drawbacks?' *Pacific Journalism Review* 26 (1): 114–22. doi.org/10.24135/pjr.v26i1.1094

Watson, Amanda HA. 2022. 'Communication, Information and the Media'. In *Papua New Guinea: Government, Economy and Society*, edited by Stephen Howes and Lekshmi N Pillai, 223–59. Canberra: ANU Press. doi.org/10.22459/PNG.2022.08

Watson, Amanda HA, Hans Adeg, Kila Aluvula and Gibson Tito. 2017. 'Telecommunication and Broadcasting Regulation in Papua New Guinea – In Conversation with the Regulator'. *Devpolicyblog*, 21 March. devpolicy.org/telecommunication-and-broadcasting-regulation-in-papua-new-guinea-in-conversation-with-the-regulator-20170321/, accessed 20 December 2022.

Watson, Amanda HA, Picky Airi and Moses Sakai. 2020. 'No Change in Mobile Internet Prices in PNG'. *Devpolicyblog*, 30 July. devpolicy.org/no-change-in-mobile-in-internet-prices-in-png-20200730/, accessed 20 December 2022.

Watson, Amanda HA, Picky Airi and Moses Sakai. 2021. 'Mobile Internet Prices in Papua New Guinea: Still No Downward Movement'. *Devpolicyblog*, 18 March. devpolicy.org/mobile-internet-prices-in-papua-new-guinea-still-no-downward-movement-20210318-1/, accessed 20 December 2022.

Watson, Amanda HA, Picky Airi and Moses Sakai. 2022. 'No Fall in Mobile Internet Prices in PNG'. *Devpolicyblog*, 19 April. devpolicy.org/no-fall-in-mobile-internet-prices-in-png-20220419/, accessed 20 December 2022.

Watson, Amanda HA, Denis Crowdy, Cameron Jackson and Heather Horst. 2020. 'Local Music Sharing Via Mobile Phones in Melanesia'. *Devpolicyblog*, 25 September. devpolicy.org/local-music-sharing-via-mobile-phones-in-melanesia-20200925/, accessed 20 December 2022.

Watson, Amanda HA and Rohan Fox. 2019. 'A Tax on Mobile Phones in PNG?' *Devpolicyblog*, 16 May. devpolicy.org/a-tax-on-mobile-phones-in-png-20190516/, accessed 17 June 2023.

Watson, Amanda HA and Beatrice Mahuru. 2017. 'Corporate Philanthropy in Papua New Guinea: In Conversation with the Digicel Foundation'. *Devpolicyblog*, 30 May. devpolicy.org/corporate-philanthropy-papua-new-guinea-conversation-digicel-foundation-20170530/, accessed 9 November 2022.

Watson, Amanda HA and Mahesh Patel. 2017. 'Telecommunications in Papua New Guinea – In Conversation with Telikom'. *Devpolicyblog*, 5 September. devpolicy.org/telecommunications-papua-new-guinea-conversation-telikom-20170905/, accessed 20 December 2022.

Watson, Amanda HA and Gary Seddon. 2017. 'Ten Years in Papua New Guinea: In Conversation with Digicel'. *Devpolicyblog*, 31 July. devpolicy.org/ten-years-papua-new-guinea-conversation-digicel-20170731/, accessed 9 November 2022.

Watson, Amanda HA and Colin Wiltshire. 2016. 'Reporting Corruption from within Papua New Guinea's Public Financial Management System'. SSGM Discussion Paper 2016/5. dpa.bellschool.anu.edu.au/sites/default/files/publications/attachments/2016-09/dp_2016_5_watson_and_wiltshire.pdf, accessed 15 July 2019.

Watt, Lucas. 2019. 'Urban *Vakavanua*: Reconciling Tradition and Urban Development'. PhD thesis, RMIT University.

Watt, Lucas. 2020. 'Fijian Infrastructural Citizenship: Spaces of Electricity Sharing and Informal Power Grids in an Informal Settlement'. *Cogent Social Sciences* 6 (1): 1719568. doi.org/10.1080/23311886.2020.1719568

Watt, Lucas. 2023. 'Magic City, Value City: The Moral Geography of Suva Fiji'. *Cogent Social Sciences* 9 (1): 2164436. doi.org/10.1080/23311886.2022.2164 436

Weber, Max. 1992 [1930]. *The Protestant Ethic and the Spirit of Capitalism*. Translated by Talcott Parsons. New York: Routledge.

Welker, Marina. 2014. *Enacting the Corporation: An American Mining Firm in Post-Authoritarian Indonesia*. Berkeley: University of California Press. doi.org/ 10.1525/9780520957954

Westbrook, Tom. 2018. 'PNG Upholds Deal with Huawei to Lay Internet Cable, Derides Counter-Offer'. *Reuters*, 26 November. www.reuters.com/article/us-papua-huawei-tech/png-upholds-deal-with-huawei-to-lay-internet-cable-derides-counter-offer-idUSKCN1NV0DR, accessed 28 November 2022.

Willans, Fiona, Jim Gure, Tereise Vaifale and 'Elenoa Veikune. 2022. 'Digicel! Topap long ples ia! An International Telecommunications Company Making Itself at Home in the Urban Landscapes of Vanuatu, Samoa and Tonga'. *TAJA: The Australian Journal of Anthropology* 33 (2): 210–46. doi.org/10.1111/taja. 12443

Willett, Ed. 2008. 'Back-to-Office Brief – Attendance at Competitive Telecommunications Industry Seminar, Port Moresby, 23 January 2008'. In *Information and Communication Technology (ICT): 'Into the Future for PNG!' Proceedings of All Day Workshop held at UPNG on 23rd January 2008*, 103–6. Port Moresby: Institute of National Affairs.

Yrteberg, Espen. 2011. 'Convergence: Essentially Confused?' *New Media & Society* 13 (3): 502–8. doi.org/10.1177/1461444810397651

Zakrzewski, Cat. 2021. '3G Shutdowns Could Leave Most Vulnerable Without a Connection'. *Washington Post,* 13 November. www.washingtonpost.com/ technology/2021/11/13/3g-service-ending-fcc/, accessed 9 November 2022.

www.ingramcontent.com/pod-product-compliance
Lightning Source LLC
Chambersburg PA
CBHW052002270326
41929CB00015B/2762